3rd International Symposium on the Science and Technology of Light Sources

Toulouse, France
18-21 April 1983

ISBN: 978-1-5108-4107-9

Printed from e-media with permission by:

Curran Associates, Inc.
57 Morehouse Lane
Red Hook, NY 12571

Some format issues inherent in the e-media version may also appear in this print version.

Printed by Curran Associates, Inc. (2017)

Foundation for the Advancement of the Science & Technology of Light Sources
FAST-LS
Belmayne House
99 Clarkehouse Road
Sheffield, United Kingdom
S10 2LN

www.fast-ls.org

Additional copies of this publication are available from:

Curran Associates, Inc.
57 Morehouse Lane
Red Hook, NY 12571 USA
Phone: 845-758-0400
Fax: 845-758-2633
Email: curran@proceedings.com
Web: www.proceedings.com

THIRD INTERNATIONAL SYMPOSIUM

ON THE SCIENCE AND TECHNOLOGY OF LIGHT SOURCES

18-21 APRIL 1983

PAUL SABATIER UNIVERSITY
TOULOUSE - FRANCE

SUMMARIES OF PAPERS

THIRD INTERNATIONAL SYMPOSIUM ON THE SCIENCE AND TECHNOLOGY OF LIGHT SOURCES

PAUL SABATIER UNIVERSITY

TOULOUSE

FRANCE

18-21 APRIL 1983

THIS IS THE THIRD SYMPOSIUM ORGANIZED TO BRING TOGETHER THOSE INVOLVED IN LAMP RESEARCH AND DEVELOPMENT THROUGHOUT THE WORLD, TO REPORT RECENT ADVANCES AND DISCUSS THE BASIC PHYSICAL AND CHEMICAL PROCESSES INVOLVED IN THE DESIGN AND OPERATION OF LIGHT SOURCES.

Secretariat : Centre de Physique Atomique
Université Paul Sabatier
118 route de Narbonne
31062 TOULOUSE CEDEX
France
Téléphone : (61) 53-08-18
Télex : UNIPSAB 521880 F

* SPONSORS *

This symposium is sponsored by :

Association Française de l'Eclairage,
Centre National de la Recherche Scientifique,
Electricité de France,
European Physical Society,
International Commission on Illumination,
Société Française de Physique,
Syndicat des Fabricants Français de Lampes Electriques,
Université Paul Sabatier.

***

* ORGANIZING COMMITTEE *

Dr M. AUBES, Paul Sabatier University, France, Treasurer,
Dr L. BARTHA, Hungarian Academy of Sciences, Hungary,
Mr E. BARTHES, Cie des Lampes, France,
Pr D. BLANC, Paul Sabatier University, France, Chairman,
Dr J. J. DAMELINCOURT, Paul Sabatier University, France, Secretary,
Dr M. KOEDAM, N. V. Philips Lighting Division, The Netherlands,
Dr H. MIZUNO, Matsushita Electronics Corps, Japan,
Dr E. W. TEN NAPEL, Twente University of Technology, The Netherlands,
Pr J. F. RAFFLE, Loughborough University of Technology, United Kingdom,
Dr S. VACQUIE, Paul Sabatier University, France,
Dr J. F. WAYMOUTH, GTE Sylvania Inc., United States,
Dr D. O. WHARMBY, Thorn Lighting Ltd, United Kingdom.

* * *

* SCIENTIFIC PROGRAMME *

MONDAY 18 APRIL 1983

8.30 Registration of participants at the symposium desk.
Opening of the exhibition.

10.00 OPENING OF THE SYMPOSIUM

- Address by Professor Daniel BANCEL, President of the Paul Sabatier University.
- Informations on the symposium by Professor Daniel BLANC, Chairman of the organizing committee.

OPENING SESSION

CHAIRMAN : Pr. D. BLANC, CENTRE DE PHYSIQUE ATOMIQUE, PAUL SABATIER UNIVERSITY, TOULOUSE, FRANCE.

OPENING PAPER

INVITED PAPER

SESSION 1

CHAIRMAN : Dr. J. F. WAYMOUTH, GTE SYLVANIA INC., DANVERS, MASS., U.S.A.

CONTRIBUTED PAPERS

13.00 LUNCH IN THE UNIVERSITY RESTAURANT.

SESSION 2

CHAIRMAN : Dr. M. A. CAYLESS, THORN EMI LIGHTING LTD, LEICESTER, U.K.

INVITED PAPER

15.15 VISIT OF THE EXHIBITION AND POSTERS.

CONTRIBUTED PAPERS

17.30 PRESENTATION OF CONTRIBUTED POSTER PAPERS

17.45 VISIT OF THE EXHIBITION AND POSTERS.

19.00 COCKTAIL RECEPTION FOR SYMPOSIUM PARTICIPANTS GIVEN BY THE ORGANIZING COMMITTEE.

20.00 BUS DEPARTURE TO THE TOWN CENTRE.

***

TUESDAY 19 APRIL 1983

8.30 VISIT OF THE EXHIBITION AND POSTERS.

SESSION 3

CHAIRMAN : Dr. A. G. JACK, N. V. PHILIPS GLOEILAMPENFABRIEKEN, LIGHTING DIVISION, EINDHOVEN, THE NETHERLANDS.

INVITED PAPER

CONTRIBUTED PAPERS

11.00 PRESENTATION OF CONTRIBUTED POSTER PAPERS

CONTRIBUTED PAPERS

16.30 VISIT OF THE EXHIBITION AND POSTERS.

SESSION 5

CHAIRMAN : Dr. D. O. WHARMBY, THORN EMI LIGHTING LTD, LEICESTER, U.K.

CONTRIBUTED PAPERS

18.30 VISIT OF THE EXHIBITION AND POSTERS.

19.00 BUS DEPARTURE TO THE TOWN CENTRE.

***

WEDNESDAY 20 APRIL 1983

8.30 VISIT OF THE EXHIBITION AND POSTERS

SESSION 6

CHAIRMAN : Dr. L. BARTHA, HUNGARIAN ACADEMY OF SCIENCES, BUDAPEST, HUNGARY.

INVITED PAPER

CONTRIBUTED PAPERS

11.00 52 - *Growth of potassium filled bubbles in doped tungsten and its relation to hot spot development and intergranular fracture.* S. GARBE and S. HANLOH, Philips GmbH Forschungslaboratorium, Aachen, Germany.....139

11.15 PRESENTATION OF CONTRIBUTED POSTER PAPERS

53 - *The reliability characteristics of tungsten halogen lamps subjected to accelerated life tests.* D. M. BARRY, Lakehead Univ., Thunder Bay, Ontario, Canada.
D. K. DAS, Marconi Company, Chelmsford, U. K.....141

54 - *New Tb^{3+} and Mn^{2+} activated phosphors and their application in "Deluxe" lamps.* J. Th. W. de HAIR and J. T. C. VAN KEMENADE, Nederlandse Philips Bedrijven B. V., Eindhoven, The Netherlands.....143

55 - *Rare earth activated niobates.* W. A. MAC ALLISTER, Westinghouse Electric Corp., Bloomfield, New Jersey, U.S.A. 145

56 - *Considerations of coiling in the design of fluorescent lamp electrodes.* E. J. COVINGTON, General Electric Co., Nela Park, Cleveland, Ohio, U.S.A.....146

57 - *Mechanism of glass darkening by a low pressure mercury discharge.* B. J. MULDER and S. VAN HEUSDEN, Philips Research Labs, Eindhoven, The Netherlands....148

11.30 VISIT OF THE EXHIBITION AND POSTERS.

SESSION 7

CHAIRMAN : E. BARTHES, TECHNICAL CENTER, COMPAGNIE DES LAMPES, PUTEAUX, FRANCE.

TRAINING SESSION

12.00 *Application aspects and quality requirements of lamps.* Prof. D. FISCHER, Univ. of Technology, Eindhoven, The Netherlands.....26

13.00 LUNCH IN THE UNIVERSITY RESTAURANT.

14.00 VISIT OF THE EXHIBITION AND POSTERS.

15.30 BUS DEPARTURE TO CARCARSONNE CITY (in front of the exhibition hall).

24.00 (nearly) BUS ARRIVAL IN THE CENTRE OF TOULOUSE.

***

THURSDAY 21 APRIL 1983

8.30 VISIT OF THE EXHIBITION AND POSTERS.

SESSION 8

CHAIRMAN : Pr. P. W. MAC MILLAN, UNIVERSITY OF WARWICK, COVENTRY, U. K.

INVITED PAPER

CONTRIBUTED PAPERS

PRESENTATION OF CONTRIBUTED POSTER PAPERS

SESSION 10

CHAIRMAN : Dr. M. KOEDAM, N. V. PHILIPS GLOEILAMPENFABRIEKEN, EINDHOVEN, THE NETHERLANDS.

TRAINING SESSION

SESSION 11

CHAIRMAN : Pr. J. F. RAFFLE, DEPARTMENT OF PHYSICS, UNIVERSITY OF TECHNOLOGY, LOUGHBOROUGH, U.K.

CONTRIBUTED PAPERS

16.45 CLOSURE OF THE SYMPOSIUM

Pr. J. F. RAFFLE, University of Technology, Loughborough, U.K.

17.00 BUS DEPARTURE TO THE AIRPORT AND TO THE TOWN CENTRE.

* * *

SUMMARIES OF PAPERS

INVITED PAPERS

I

CONFERENCES INVITEES

RESUMES DES COMMUNICATIONS

WHAT DO WE SEE, DO WE REALLY SEE BY E. BARTHES

VICE PRESIDENT OF CNFE
CHAIRMAN OF CIE T.C. 3.5. LUMINOUS ENVIRONMENT

CENTRE TECHNIQUE MAZDA
11 Quai de Dion Bouton
92818 PUTEAUX CEDEX

Vision as stated in the Encyclopedia Britannica, is the function in which the organ known as the eye and its more immediate nervous connections in the brain are crucially involved in a series of reactions to the form of radiant energy called light, but the behaviour of the eye is by no means sufficient to account for vision.

If the light acts uniformly over the entire retina we are just aware of light without reference to an external object.

Eventually we are sensitive to the fact that the light is coloured if the coulour adaptation is not perfect.

To have a real vision of the external world, a kind of copy of the external objects, in terms of distribution of light on the retina , must be obtained due to the refractive structures of the eye which focus the luminous rays.

Up to that point the spectral radiometric knowledge of the light falling on a given point on the retina is enough to understand vision.

Practice already indicates that it is not true. Just consider the colour of a thread and the colour of the cloth obtained with it. The results are certainly different. And what about the simple effects of the nature of the fabric : silk or tweed ?

It is clear that to be understood by the brain the radiometric information must be coded by specific channels, the study of which is fruitful.

Colour contrast between detail and background are the basic information but many other cues are available to understand objects.

Real knowledge of the needed visual information and of the redundant one will be obtained by that type of studies.

How can we utilize that knowledge to design new types of lamps ?

Formerly the spectral distribution looked for in the range of white discharge lamps was continuous, trying to reproduce natural effects such as incandescence or daylight.

Then the three band type was created where the goal was to satisfy as well as possible the physiology of the retina.

Each band is in accordance with one of the spectral sensitivity of the light receptors.

This solution permits to have simultaneously a good colour rendering and a high efficacy.

But now we can think about overpassing the physiological aspects to be concerned with the visual information the brain is able to treat and understand.

Many studies have been done on colour preferencies but never with great success.

Clearly skin tone good colour rendering is of prime importance, but with experience it appears that lamps with excessive red content such as soft white are not prefered to cool white with good colour rendering.

So probably a natural colour rendering is necessary. By natural we mean that dramatic changes of colour must be avoided on objects which are familiar and also that the classification of different shades or tints should remain approximatively the same when passing from daylight to electric light.

Colour constancy is a "must", but if colour constancy can be achieved without using continuous spectrum of light, due to the brain information treatment and by using a combination of monochromatic radiations conveniently choosen to produce the good brain answer then a highest efficacy will be obtained.

Precise psycho physical tests should be effected to get new knowledge in that field.

COMPACT LOW POWER DISCHARGE LAMPS FOR GENERAL LIGHTING

M A Cayless
Head of Lamp Research, THORN EMI Lighting Limited, Leicester, England

For many years people have searched for a discharge lamp that could replace the ordinary incandescent lamp in the home, and other interiors of a more intimate character. One of the main motives has always been the comparative inefficiency of the GLS lamp. During the last decade the motivation to conserve energy has intensified this effort and, in some places at least, substantial funds have been available to pursue this goal.

Such a replacement needs to fulfil several requirements. It must: (a) be more efficient than a GLS lamp; (b) have a relatively low power; (c) have good colour properties; (d) be simple, safe and interchangeable with a GLS lamp; (e) be reasonably cheap, and have a long life. Other desirable features are: (f) instant start and restart; (g) effective dimming capability. A tall order !

Developments have proceeded along two distinct lines:

<u>Low pressure lamps</u>. These are all miniature variants of the ordinary fluorescent lamp, with the discharge path convoluted or partitioned in some way to give a more compact and congenial geometry. To give a convenient light output with attractive energy saving, power consumed is less than 40W, or even 20W.

<u>High pressure lamps</u>. These are mostly miniature versions of metal halide discharge lamps, less than 150W, and as low as 30W.

In the former, the main objective is a compact enough geometry, in the latter, a low enough power.

To be an attractive proposition, the control gear must be compact, and linked with the lamps so that the combined unit can be exchanged for a GLS lamp without more ado.

with the discharge into a single "bulb". When this fails, the whole is thrown away and replaced. The other approach is to make the two components separable, but fitting together in some more or less ingenious way. Various compromises have been used, such as making the starter integral and the ballast separable.

For compactness and lightness electronic ballasts have an attraction, operating at a higher frequency than mains. This also helps keep the end losses low, always a problem at low powers, and improves prospects for instant starting and dimming capabilities desirable for integrated energy mamangement systems. They tend, however, to be expensive and hard to keep cool enough.

At high enough frequencies, r.f. or microwave, electrodes can be dispensed with, and end losses disappear. On the other hand, circuit complexity and losses increase, and there are problems of r.f. interference. Nevertheless several serious proposals have been made and the attraction of avoiding electrodes and their problems cannot be gainsaid.

Low pressure lamps

Low wattage fluorescent lamps have been available for many years, but efficacy and maintenance tend to fall off at low powers. Further, the straight tube geometry is found, rightly or wrongly, to be aesthetically unattractive. Circular and U-shaped lamps have been about for a long time, and in some countries are popular, but not in others. There has been great exploitation in the last few years of other compact geometries. Tubes have been bent into helices, D-shapes, O-shapes, bent U's or linked bars; an alternative has been to partition a wider tube or bulb into a convoluted discharge path: think of a topological variant of a long cylinder, and someone has tried it !

The reason for this world-wide burst of enthusiasm, apart from the fashionable energy conservation, is a technical one. The discharge intensity on the phosphor coating is greater than in more conventional lamps and ordinary phosphors decay rapidly. It is only the advent of newer phosphors, based on rare earth activators which have made small, compact but bright long life lamps possible.

The same problem has hitherto restricted the development of electrode-less versions. One widely publicised development had a GLS-shaped bulb surrounding an exciter coil powered from a unit in the base. The loading on this bulb would cause rapid deterioration of a conventional halophosphate, but becomes much more acceptable with the newer phosphors. The principal remaining problems are associated with the circuitry, heat dissipation and cost, and with r.f. interference.

High pressure lamps

The demands of reasonable efficacy and colour properties have restricted these largely to metal halide discharges. It becomes increasingly difficult to make these at powers below about 100W. Much of the power is dissipated simply in keeping the arc tube and electrodes hot enough, and the high surface area to volume ratio makes the lamp chemistry more critical. Impurities are more difficult to exclude, and surface reactions tend to alter the contents and hence the light produced. For this reason much of the activity has been in the 70-150W region, though powers as low as 30W have been explored.

There has been a noticeable tendency to exploit the properties of tin halides, which can be advantageous at low powers. They also have the property of forming voltatile complexes with other halides which would otherwise be unuseable. In some cases there is deliberate excitation of radiation from molecular species, such as tin monochloride, which effectively produce a continuum. These are sometimes referred to as molecular arcs. In others the halide is simply the means of carrying the metal atoms as a vapour which is dissociated in the discharge, to produce largely atomic line radiation.

To achieve the necessary power densities at low power, the arc tube must be small, the arc length short, and the mercury vapour pressure high, making severe demands on conventional silica. One solution is to use a translucent alumina arc tube, as in the high pressure sodium lamp, but alumina itself is not completely inert, and the halides tend to attack the end seals and electrode leads. One manufacturer uses a conducting cermet end instead of a metallic through lead, which eases these problems. Another approach is to use a silica tube with an isothermal geometry, ellipsoidal or spherical. By paying particular attention to the geometry of the end seals, so that heat loss is minimised and light transmission maximised, reasonably efficient lamps have been made at the lowest powers so far achieved, 30W or so.

A particular problem with high pressure discharges is the finite warm-up time, and the restrike time after extinction. Electronic control circuits operating at high frequency are usually prescribed, but when combined into a single plug-in unit with the bulb, both geometrical and thermal considerations necessitate a much larger than normal base and specially developed components for critical circuit elements. The very low power types just mentioned have been embodied in a version combined with an incandescent filament which provides instant light when switched on.

Electrodeless high frequency discharges are conceptually even more attractive in the high pressure region: a small sphere or tube without complications of electrode or lead-throughs has obvious advantages. Such sources, excited at around 100 MHz or microwave frequencies have been used for many years for special purposes, such as spectroscopy, but for general use there are still many problems, apart from those of adequate generation efficiency and radiation screening.

Problems and prospects

The chief problem areas at present are:

(1) Compact, light, efficient, robust and cheap control circuits;

(2) Phosphors or arc tubes resistant to the discharge power loading;

(3) Reduction of electromagnetic or acoustic interference.

All these are easier at present with the low pressure forms. (1) is more acute when the control unit is regarded as integral with the discharge tube. Electronic control units appear to offer most advantages except that of cost, and ideal forms seem to await just one more major electronic component development. Further improvements in (2) may confidently be expected. Given effective solutions to these then the necessary incentive to overcome (3) will surely be found.

Low pressure versions are already available in variety, and high pressure forms are on the way: these will complement each other in appropriate applications. We may suppose with confidence that much more subtle devices will develop form them in the next decade or two. We may expect to see them displacing incandescent lamps in an increasing number of applications, especially in conjunction with electronically controlled lighting and energy management schemes: just how many and how fast is going to depend as much on economics as on technology. For some considerable time individual lamps are going to be more expensive than GLS lamps, and the economics of pay-back calculations are going to matter less than the psychology of sales-appeal versus unti price.

High-Pressure Sodium Lamps

H. Akutsu
Matsushita Electronics Corp., Light Division
Osaka, Japan

This review is intended to clarify the present state of the high-pressure sodium (HPS) lamp, and eventually indicate its future prospects. The recent activities give us an impression that there is a new trend in the HPS lamp that it steps forward from the first technological maturity to the next evolution.

1. Analyses on Physical Properties of HPS Discharge

The physical properties of the HPS discharge have been generally well understood through theoretical analyses and reliable experiments. Here, it is noted that the former analyses have a significant contribution in offering an outline of the HPS discharge behavior. This owes primarily to that:

(a) The assumption of local thermal equilibrium can be reasonably applied to the central portion of the HPS discharge, except near-the-tube-wall.

(b) Relevant data on material functions of the HPS discharge have been available, although selection of them is difficult. A definitive measured data on the sodium-mercury pressure is presented recently.

(c) Accurate plasma temperature profiles along the tube radius were obtainable, most satisfactorily through the spectroscopic measurements; almost parabolic profile with axis and wall temperatures of around 4,000 and 1,500K, respectively.

The HPS lamp dissipates more than half of the input power as the radiation; sodium D-lines ~22%, nonresonance lines and continuum each ~15%, and shows the peculiar radiant properties with the strong reversal around the D-lines. A first successful analysis was given by calculating the radiation transfer equation on the basis of the resonance broadening for the D-lines. This clearly demonstrated the significant role of the self-absorption on the reversal of alkaline resonance lines. The calculation including additional strongest nonresonance lines and partial continuum has been extended so far, providing an appreciable agreement with measured visible radiance except the far wings of the D-lines. Recent investigations on details of the radiance including continuum in the infrared region clarified or suggested some roles of each of Na_2 molecule, Na-Hg and Na-Xe interactions, but did not give a full understanding about such details yet. Also reported is a possibility that the increasing Xe buffer gas pressure results in the less convex temperature profile, offering a kind of spectral broadening effect. In order to obtain definitive justifications on some of these respects, more work is still needed.

The various plasma properties such as temperature profile, field strength, radiant efficiency and conduction loss, have been calculated, mostly using the energy-balance equation together with the ohm's low. These offered in general only qualitative agreement with measured results, because of the uncertainties in the calculations especially of the radiant flux. Neverthless, what is remarkable is that the calculations not only revealed basic mechanisms of the HPS discharge but also gave an elementary explanation about the relationship between the discharge parameters and properties: The HPS discharge features the self-absorption in the outer cooler region of the arc tube, which decreases the radiation from the central portion while presenting the convex temperature profile. Here, to cope with the difficulties in the calculations, also introduced were the approximation models employing a position-dependent escape factor, a semi-empirical formula of the net emission coefficient and a two-step-temperature profile, respectively. Recently with the two-step-temperature model, an interesting speculation to improve the HPS lamp properties is presented.

2. Recent Technical Trends of HPS Lamps

In view of the high operational reliability as well as the high efficacy, present day commercial standard HPS lamps (CRI 25 and CCT 2,050K) may be considered to have almost reached a first technological maturity.

The standard HPS lamp with a low pressure Xe starting gas has been designed for maximum efficacy in consistence with requirements. This is based on almost empirical design forms, which provide an appropriate combination of various design parameters of the arc tube: The general design range is the sodium pressure 20-80 Torr, the amalgam composition of Na 70-80 mole% and the tube wall loading ~20 W/cm^2. In addition, various improvements on optical properties of the polycrystalline alumina ceramic (PCA) tube have been done, making a significant contribution in increasing the efficacy. Thus, the present day standard HPS lamp employing such refined PCA tubes offers ~20% higher efficacy (~125 lm/W at 400W version) than that of the initial lamps. This seems to have approached the attainable maximum efficacy for the PCA tube with the low pressure Xe gas.

It was anticipated already in the early stage of the HPS lamp development that a high pressure Xe buffer gas could improve the efficacy because of its lower thermal conductivity. Recently, based on this concept, a new standard lamp with the additional high pressure Xe gas (200-350 Torr) has been materialized, providing still ~12% higher efficacy. The new lamp has shown a high reliability almost equal to that of the low pressure Xe-filled lamp, although the lamp requires a special ignition wire on the arc tube.

A notable trend in recent developments is the pursuit of improvement in the color rendition. It is well-known that, as the sodium pressure is increased, radiant spectrum from the HPS discharge becomes broadened over the whole visible region, thereby improved color rendering. Also confirmed from recent works is that increases both of inner tube diameter and Xe buffer gas pressure are effective for additional spectral broadening.

Based on the broadening effects of these parameters, a color-improved HPS lamp (CRI 60 and CCT 2,150K) and a super HPS lamp (CRI 85/78 and CCT 2,500/2,800K) have been introduced during recent years. Basic design of these lamps is almost the same as that of the standard HPS lamp. Main differences are such that, to increase the sodium pressure, thermal protective layers of tantalum or niobium plate are applied at both ends of the arc tube, and, especially in design of the super HPS lamp, the broadening effect of larger tube diameter is fully utilized.

Such color improvement is attainable at the sacrifice of the efficacy, and, more time may be needed to bring up these lamps to a complete device. Neverthless, the new type lamps will make a way for the HPS lamp to general indoor lighting services.

Contrary to the high reliability of the standard HPS lamp, there have been occurring some problems on life performance of the new type HPS lamps because of the more critical operational conditions.

Since the invention of the HPS lamp, improvements on construction of the arc tube have been steadily carried out to achieve reliable vacuum-tight sealing, better life performance and simpler tube as well. In recent years, two types of tube constructions, monolithic and alumina disc-capped tubes, seem to take a main position. Also noted is the introduction of a sophisticated construction with lead-in niobium rod. This would reveal its significance especially in the smaller wattage types.

The operational wall temperature of the PCA tube is limited below ~1,500K to prevent sublimation of Al (dominant), AlO and Al_2O from the tube wall, although the higher wall temperature offers the higher efficacy.

Sodium loss from the arc tube during life results in rising of the lamp voltage as well as shifting of the lamp color, therefore, causing a short lamp life. The sodium loss in the PCA tube was a rather complicated problem, and became more remarkable in the new type HPS lamps. The skillful analyses, however, have been gradually revealing various processes of the sodium loss such as leakage through the sealing glass, diffusion/electrolysis through the tube wall and reaction with the tube surface. As the results, there appear suggestions for solution or principle to suppress each of the loss processes that are effective even for the new type lamps. Against the sodium leakage, selection of the sealing glass is important, and also the intermediate layers on the niobium tube (Si and W) are effective. Furthermore, it should be noted that, not only depending, on the tube wall temperature, other loss processes (diffusion and reaction) also may be triggered or accelerated by the impurities, in particular calcium, doped in the PCA tube and released from the electrode and the sealing glass.

The exhaustion of the emitter due to both sputtering and evaporation is another problem, which also brings about the lamp voltage rise as well as the questionable impurities. The influential factors on this problem have been almost experimentally surveyed so far, although they are difficult to be regarded as a conclusive analysis. The new emitter materials and the increasing Xe gas pressure can reduce to some extent the emitter exhaustion. In addition, a remarkable effect is attainable with the renewed concept in electrode design, which presents much shorter glow-to-arc transition time and high stability of the cathode spot during long life.

3. Future Prospects of HPS Lamps

During the coming years, energy scarcity will keep on strengthening the position of the standard and color-improved HPS lamps in outdoor lighting and industrial lighting as well. The new higher efficacy lamp with the high pressure Xe buffer gas would take, in view of its high reliability, a significant portion among the total HPS lamps. Also expected is that the super HPS lamp would find application fields of its own in interior lighting, especially for commercial lighting.

Further significant improvement in the efficacy of the HPS lamp requires introduction of either a new tube material resistive against higher temperature or an evolutional concept in lamp design. As a new possible mateiral, single crystal alumina (sapphire) has been tested, and recently sintered yttria with a lower total emissivity is being investigated. The sapphire might be feasible, although having still questions in its reliability and cost. In respect of a lamp design, proposed are rather speculative concepts such as a thermal reflecting layer on the outer-bulb, a highly loaded lamp with a filling gas in the outer-bulb and a larger diameter arc tube combined with some heating means.

A goal of the super HPS lamp would be to develop the compact type as an energy-saving substitute for incandescent lamps. This offers a great feature of the excellent color rendering properties almost comparable to those of incandescent lamps. It may be technically feasible to realize the compact lamp with an efficacy over 40 lm/W, CRI 85 and CCT 2,500K with 50W version.

Another future subject may be how to develop a better operational circuit including starter device. The HPS lamp has in its nature relatively poor operational characteristics, basically because of the operation at the saturated sodium-mercury pressure. In the future, advancement in solid-state technology might pave a way to a sophisticated electronic circuit offering a constant lamp voltage operation or a high frequency operation accompanied even with an interesting pulse-current mode.

LAMPS AND CIRCUITS

H. Lange
OSRAM Research, 8000 München 90, Germany

1. Introduction

The first electrical supply sytems originated from the invention of the incandescent lamp. The choice of a line frequency of 50 or 60 Hz resulted finally from the need to reduce light flickering. With the introduction of the discharge lamps about fifty years later, these lamps had to be designed in accordance with the operating requirements of the different supply systems existing all over the world. Since the early days of discharge lamps it has been known that the operation at higher frequencies increases lamp efficacy. The progress made in the field of semi-conductors enabled us to develop reliable and economic ballasts for fluorescent and HID-lamps. In comparison with incandescent lamps, however, we are dealing with a rather complex system of lamp, supply network and ballast, which influences the performance of the lamp. The appropriate operating frequency range is between 20 and 70 kHz, because of acoustic disturbance with lower frequencies and limits in generating higher frequencies. New experiences with the semiconductor ballasts and the results will be discussed for different lamp types.

2. Fluorescent Lamps

The greatest progress so far has been made with high-frequency ballasts for fluorescent lamps. The technical aim is directed towards substantial reduction in ballast and lamp losses.

The essential elements are:

* high frequency generator (35 kHz)
* radio frequency interference suppressor and filtering device for the reduction of harmonic frequencies in the line current
* the current limiting circuits (up to 4 lamps).

Approximately 50 to 60 electronic components are required for a circuit. The use of a choking coil as a harmonic filter instead of an electronic circuit reduces considerably the number of components and, therefore, the cost of a ballast.

In addition to the overall efficacy of lamp and ballast, the following improvements concerning lamp operation will be obtained:

* no flicker at the electrodes
* no stroboscopic effect
* instant start (less than 1 second)

Which are the reasons for the improvement of efficacy with high frequency operation? Comparing the losses of different ballasts, the measurements have shown a reduction of almost 60 % compared with standard ballasts. These have, for instance in the case of 5-ft-lamp, a power consumption of approximately 13 W. The power consumption of an electronic ballast is only 5.5 W. In the lamp an additional improvement occurs. Due to the polarity changes, the electrodes remain emissive, and a sufficient amount of electrons is always available for the discharge process in the following half cycle. It is assumed that the necessary electrons for the discharge are directly supplied by the electrical field. Concerning the total power consumption there are lower electrode losses but a higher power consumption of the positive column. The UV-efficiency increases only slightly.

3. HID-Lamps

In principle, similar circuits as for fluorescent lamps can be used for HID-lamps. However, acoustic resonance phenomena will be noticed which are not existing in fluorescent lamps. They are caused by periodical pressure fluctuations in the discharge which depend mainly upon temperature, the physical dimensions of the lamp and the filling gas. In the case of resonance the fluctuations can be so excessive that the arc tears off and the discharge is extinguished. In case of the HPS-lamps, they can even destroy the PCA-arc tube. The resonance effect takes place in a range between 1 and 150 kHz. The resonance frequencies depend on design date and manufacturing process. Therefore, the operating frequency of the ballast must be controlled thoroughly in contrrary to the ballast for fluorescent lamps.

As in the case of fluorescent lamps, important lamp parameter are changed. Outstanding is here the change in arc shape. At HF-operation, the arc is straighter than at lower frequencies, apparently because of the fast period changes do not allow sufficient time for the electrons to deviate from a straight path.

4. Halogen Lamps

For incandescent lamps new possibilities are opened by using electronic circuits. At present, the design of high voltage (220 V) halogen lamps is limited to 150 W. The wire quality of today, particularly the accuracy of the wire diameter, is not sufficient for lower wattage halogen lamps. Here, electronic circuits are well suited for the design of "electronic transformers" to supply the incandescent lamps with a lower voltage in order to guarantee a perfect functioning of the halogen process. Savings in size, weight and efficacy will be the advantages of these devices.

5. Compact Fluorescent Lamps

Small compact fluorescent lamps are offered now to replace incandescent lamps. In these lamps the ballast must be incorporated. Electronic circuits present themselves as ballasts because of their small size and weight. Technically we still have to improve the thermal stability of some electronic components and to reduce the costs of the circuits, because this lamp type has to stand up in competition with incandescent lamps. For the design of the shape of the discharge, the use of high frequency is of great importance.

6. Influence of Supply Network to Lamps

More and more ripple control signals are used for transmitting commands over existing electrical supply networks. The signal can influence the light emission of lamps resulting in more or less flickering. The possibilities to reduce this effect by means of conventional lamp and ballast design are negligible. Only semi-conductor ballasts can be designed to avoid these interferences completely.

7. Summary and Outlook

Semi-conductor ballasts open new possibilities in lamp technology:

* Fluorescent lamps:
  Considerable energy saving and product advantages; it is expected that in the near future this ballast design will succeed and a reduced pay-back-time should be available.

* HID-Lamps:
  In addition to the energy saving, the reduction of weight as well as the possibility to change the lamp-design and parameters will be extremely important.

* Compact Fluorescent Lamps:
  The break-through to mass production is expected for the second half of the 1980's, if the electronic components can be brought to a sufficient level in technology and costs.

With the operation at high frequency ballasts the lamp designer has the chance to overcome the differences in lamp construction, caused by the different line voltages existing all over the world. This could originate a new period of lamp development.

DIAGNOSTIC METHODS IN LAMP RESEARCH

W.J. van den Hoek

Philips Lighting Division, 5600 MD Eindhoven, The Netherlands

1. Introduction

Lamp diagnostics in general involves the characterization of lamps in terms of basic physical and chemical quantities. Such studies cover a broad field. The present review, however, will be restricted mainly to recent developments in the field of spectroscopic plasma diagnostics of gas discharge lamps. Progress in classical spectroscopic methods has been made mainly in the field of high-pressure discharges. Progress in laser spectroscopic methods covers low-pressure discharges as well.

2. Quantities to be measured

The Table below lists the basic quantities to be evaluated from diagnostic measurements on discharge lamps. In the majority of cases plasmas in discharge lamps are strongly inhomogeneous and vary periodically (50-60 Hz) with time. An important condition for plasma diagnostics is the availability of basic atomic and molecular data such as transition probabilities, line-broadening parameters etc.

BASIC PLASMA QUANTITIES

- electron temperature	T_e	(K)
- gas temperature	T_g	(K)
- electron density	n_e	(m^{-3})
- chemical plasma composition (atom, ion and molecular densities)	n_x	(m^{-3})
- excited state densities	n_x^*	(m^{-3})
- characteristic relaxation times for plasma processes	τ	(s)
- impurity levels	n_x	(m^{-3})

3. The importance of diagnostics

Three aspects of diagnostics may be distinguished with respect to lamp research and development.

a. Model studies. Characterization of the plasma is a prerequisite for the construction of physical-mathematical models of gas discharge lamps.

b. Fast methods of plasma characterization. In practical lamp development fast characterization of the discharge is often needed in order to determine the course of further work. This requires quick and simple techniques.

c. Impurity diagnostics. Impurities often play a decisive role in lamp behaviour during life, so detection and quantification of impurity species is important.

4. Passive and active diagnostics

Spectroscopic plasma diagnostics can be divided into two categories (Fig. 1). In passive diagnostics the radiation that escapes is simply measured (emission spectroscopy). The interpretation of emission measurements, however, is not simple. The measured intensity always represents an integral of local emission and absorption over the optical path in the discharge. A particular problem is the study of the outer regions (low-temperature regions in arcs).

Fig. 1
Schematic indication of passive (top) and active (bottom) diagnostics (D = detector)

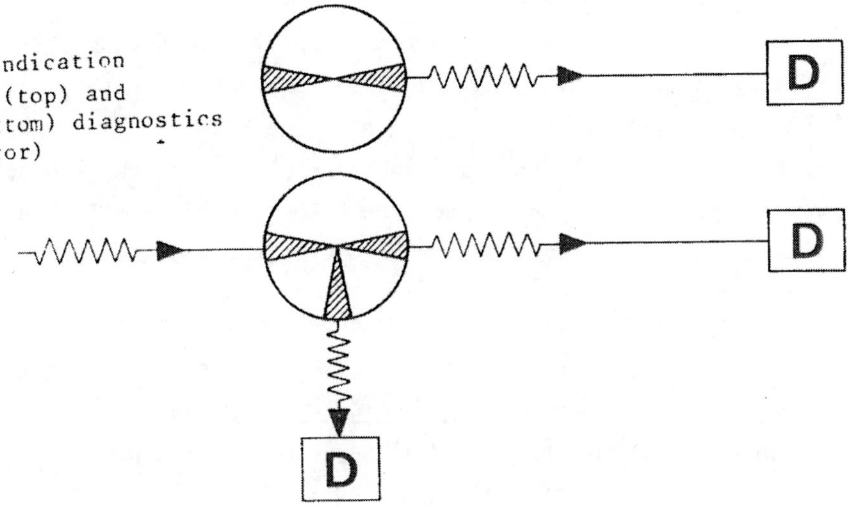

In active diagnostics the plasma is probed by a laser beam and the results are observed of the interaction between the laser light and the plasma. Here a variety of experiments are possible. By appropriate combination of probing and observation local plasma properties can be measured. Besides the outer regions can be studied. Absorption measurements using conventional light sources also belong to the category of active diagnostics.

5. Recent developments

5.1. Emission spectroscopy

Notable trends in emission spectroscopy are

a. A better understanding of the spectra of high-pressure discharges. For metal halide lamps this has resulted in the development of fast methods for estimating additive densities based on the separation of self-reversal peaks (Δλ-methods).

b. The use of computers coupled to conventional optical spectrometers for measurement control, data handling and computation (e.g. integral transformations). This facilitates more complete measurements of inhomogeneous systems and also enables short-term and long-term lamp life studies to be carried out.

c. The use of optical multichannel analyzers coupled to a computer system. These devices enable the registration of a complete emission spectrum in a fraction of a second. Phenomena like lamp ignition and run-up, instabilities etc. can therefore be studied more easily.

5.2. Laser spectroscopy

Lasers are ideal light sources for active lamp diagnostics because of - among other things - the high light intensities with small spectral bandwidths, the geometrical properties of the beam and the possibility of continous wavelength tuning (dye lasers). A drawback is often the limited wavelength range. Lamp diagnostics using lasers is still in the exploratory phase. Techniques developed or used over the last few years are:

a. Light scattering. Local particle densities from the arc axis up to the wall have been determined in metal halide arcs by electronic Raman scattering. Near-wing Rayleigh scattering has been used for local Na density determinations (metal halide arcs, low-pressure Na discharge).

b. Optogalvanic spectroscopy. With this technique impedance changes caused by irradiation of the discharge with a dye laser beam are detected. Characteristic times for energy relaxation in Hg and Na arcs have been studied.

c. Laser-induced fluorescence. In an Na arc local electron densities have been measured from Stark shifts in the laser-induced fluorescence using broad band excitation with a dye laser.

d. Doppler-free two-photon spectroscopy has been used to determine local electron densities from Stark broadening and shifts (low-pressure Na discharge).

e. Interferometry. Excited state densities in a low-pressure Hg discharge have been determined with the hook method using a pulsed dye laser.

f. Absorption spectroscopy. The advantages of lasers over conventional light sources are obvious with absorption experiments.

6. Future development

Emission spectroscopy will undoubtedly remain the basic tool in lamp diagnostics in the future. It can be expected, however, that continuing exploration of the application of lasers will result in standard techniques for lamp characterization. It may be expected that laser spectroscopy will provide insight into more difficult problems such as the chemistry of the low-temperature region in metal halide arcs, impurity levels and chemical transport processes.

SOME ASPECTS OF LAMP MATERIALS

P.W. McMillan
Department of Physics
University of Warwick
Coventry, CV4 7AL
ENGLAND

The topic of lamp materials covers a very wide field including metals for filaments, electrodes and lead-throughs; envelope materials and other non-metallic structural components; coatings for light dispersion and other purposes; phosphors and in more recent developments, semiconductor materials for light-emitting diodes and related devices.

Within the scope of the present paper it is not possible or desirable to attempt a full review of this broad field and therefore the discussion will largely be confined to envelope and non-metallic structural materials with fairly brief reference to other materials where these are appropriate to the main theme. The materials chiefly of interest are glasses, ceramics and glass-ceramics.

A lamp envelope comprises both metallic and non-metallic materials and an essential feature is that the materials selected are compatible in a number of respects. The formation of seals with the necessary mechanical integrity, which are vacuum leak-tight and which will withstand the thermal cycling imposed during processing and operation, generally requires fairly close matching of thermal expansion coefficients over a relatively wide temperature range. A further important aspect is that the metal chosen must be capable of being wetted by the sealing material (glass, glass-ceramic or braze metal) to form a seal that is impermeable to gases under all conditions of operation and use. These factors strongly influence the choice of systems that can be employed. The paper will discuss the characteristics of metals that can be utilised in the light of the foregoing characteristics.

Considering the non-metallic components of lamp envelopes the material selected should fulfil the following requirements. It must:-

(i) be transparent to visible radiation or at the least, possess a high degree of translucency.

b) possess an adequate mechanical strength

c) be impermeable to gases

d) possess long term physical and chemical stability under the operating conditions

e) be resistant to interactions with gases or vapours contained in the lamp which may cause loss of transparency or other undesirable effects.

f) possess the necessary electrical insulating characteristics

g) be capable of sealing to the chosen metal by economic processes.

The paper will analyse and discuss the characteristics of envelope materials in the light of the foregoing requirements.

Glasses have long been established as the most important class of lamp envelope materials. They combine many of the desirable physical characteristics required in an envelope materia. The valuable properties of glass in this context derive from its non-crystalline or liquid-like structure and the relationships between glass structure and important physical properties will be briefly reviewed. Important aspects of glass behaviour that bear upon its suitability as an envelope material will also be discussed. These include stabilisation effects in the transformation range and glass-in-glass phase separation. The latter can lead to significant changes in physical properties such as electrical characteristics and under certain conditions to impairment of chemical stability. Other possibl limitations on the use of glasses for certain applications include the effects of imperfect matching of thermal expansion coefficient to that of the chosen metal. The residual stresses generated in the glass and the effects of thermal cycling on these can result in the failure of seals, sometimes after a protracted period, owing to the occurrence of static fatigue. The fundamental causes of this phenomenon will be discussed. A further limitation upon the use of glass in some applications relates to the occurrence of undesirable interactions with gases or vapours which are present in the lamp filling. Reactions with alkali metal vapours can be especially damaging and this problem will be discussed in the light of recent research at the University of Warwick which has provided information concerning the fundamental mechanisms of the process of attack.

To complete the discussion of glasses, a brief survey will be given of the commercially available types that are important in the manufacture of lamps.

Ceramics are polycrystalline solids which can be produced from minerals or from pure oxide starting materials. For lamp envelopes where translucency or transparency are necessary, only pure oxide ceramics are suitable. The achievement of the necessary optical characteristics requires careful choice of starting material, control of grain growth and elimination of porosity during the sintering process. These aspects will be discussed in the light of the fundamental processes involved.

Possible transparent/translucent oxide ceramics will be reviewed with emphasis on translucent alumina which is firmly established as an envelope material for high-pressure discharge lamps.

The advantages of ceramics as compared with glasses for certain applications will be discussed. These rest on the high chemical stability which renders them (especially pure alumina) highly resistant to interaction with alkali metals and halogen gases at elevated temperatures. Ceramics generally possess higher mechanical strengths than glasses and the reasons for this will be abalysed.

The chemical inertness of pure oxide ceramics and their refractoriness mean that sealing them to metals results in some difficulties. It is necessary to employ an intermediate bond which can be a braze metal used in conjunction with a suitable premetallising process for the ceramic. Other possibilities exist however, of utilising refractory glasses as a thin bonding layer between the ceramic and the metal and the use of crystallisable glasses (glass-ceramics) is a further possibility.

Glass-ceramics, which are made by the controlled crystallisation of special glasses, possess a range of properties that render them potentially valuable as materials for lamp construction though they have not so far achieved widespread application. The method of preparation of glass-ceramics and the means by which controlled nucleation and crystal growth are achieved in order to generate the desired very fine-grained microstructure will be described. It will be shown that by suitable choice of parent glass composition and of the heat-treatment cycle, high strength glass-ceramics having a wide range of valuable physical properties can be produced. In particular, materials matched in thermal expansion coefficient to a large number of useful metals are possible.

Glass-ceramics may be regarded as combining a number of the useful characteristics of glasses and conventional ceramics. Thus, using suitable techniques, they can be sealed directly to metals without the use of intermediate bonds, but at the same time they possess high mechanical strengths and good refractoriness. Methods of sealing glass-ceramics to various metals will be described to illustrate the versatility of the materials.

For use as a lamp envelope, transparent or translucent glass-ceramics are required and the principles underlying the development of such materials will be reviewed with reference to practical examples. These will include glass-ceramics matched in thermal expansion coefficient to refractory metals. It will be shown that glass-ceramics possess good chemical stability and that translucent materials which are resistant to attack by sodium metal vapour can be derived.

In addition to their potential as lamp components, glass-ceramics have considerable promise as sealing materials for ceramic to metal joining for critical applications and the materials and techniques available will be outlined.

In the concluding section of the paper significant recent developments in inorganic non-metallic materials processing that may be of interest and importance in future developments of lamp envelope technology will be reviewed. These will include the production of glasses and glass-ceramics by the sol gel route, surface modification techniques for glasses and the production of vitreous materials by extremely rapid quenching techniques.

APPLICATION ASPECTS AND QUALITY REQUIREMENTS OF LAMPS

Prof.Dr. D. Fischer
University of Technology, Eindhoven
Chairman of CIE TC-4.1 'Interior Lighting'

Introduction

There has always been a mutual influence and feedback between lamp development on the one hand and lighting application on the other. Although there are various influences on lighting application at work, such as changing trends in architecture and interior design, results of investigations into quality criteria of lighting, cost and energy consumption considerations, and the vast experience gained in practical lighting design and engineering, the main impetus for reviewing the principles of lighting application has always come from the lamp development side. This was the case some hundred years ago when the first serviceable electric lamp was presented; it was the case when the modern discharge lamps came on the market, and again recently when the compact fluorescent lamps were introduced.

With respect to these revolutionary developments of new lamps based on new principles or technologies, the contribution from the application side was practically nil. However, there is also the other extreme where certain lamps, e.g. automobile headlamps and projection lamps, are developed or adapted to suit specific applications, in which case close co-operation between lamp developers and application engineers is always given. It would seem more appropriate in this paper, therefore, to deal with general lighting practice whilst at the same time keeping an eye on the relevant lamp properties.

International recommendations

There are two documents on lighting criteria and recommendations for interior lighting being prepared by international organisations.

These are:

1) Final draft (November 1981) of ISO Standard 'Principles of visual ergonomics to be applied to the lighting of work systems'

2) Draft (November 1982) of Publication CIE No. 29 (TC-4.1) 'Guide on interior lighting'.

The main criteria being dealt with in these codes are illuminance, glare limitation, colour qualities of light sources, and contrast rendering. The recommendations are based on research and practical experience, and are aimed at achieving good visual performance and visual satisfaction and an economic use of energy.

In order to meet these requirements and to create a satisfactory visual environment, the designer has to correctly select lighting systems, lamps and luminaires. Where the lamps are concerned, this selection will always be made bearing in mind their suitability for the space being lighted and the requirements of the occupants, as well as their suitability for the lighting system and type of luminaire chosen for the installation.

Consider, in turn, the above-mentioned lighting criteria and how each is dependent on certain lamp characteristics.

Illuminance

Since the early seventies the lighting levels recommended within Europe have remained practically unchanged, and higher illuminances than those recommended are rarely installed. The requirement for lamp development was, and is therefore, that in order to enable existing installations to make use of the increased luminous efficacy of lamps, the luminous flux of such lamps has to be maintained and the wattage reduced accordingly.

When calculating the number of lamps and luminaires needed for a specific space and a given illuminance, the designer has to employ a light loss factor to allow for the fall in the luminous flux of the lamps and the depreciation caused by dirt on luminaires and room surfaces. The smaller the light loss, the fewer is the number of lamps that need to be installed to maintain the recommended lighting level. It is therefore an obvious quality requirement to keep the lamp lumen depreciation as small as possible.

The need to conserve energy and the general desire to create a more comfortable and cosy atmosphere in offices, together with the demand for a certain individual control of the environment, has led to a revived interest in task lighting. However, as experiments in a simulated office have shown, general lighting ('ambient lighting') should represent at least one third but preferably 50 per cent of the total illuminance in order to create an acceptable visual environment.

This means that, for a typical 500 lux office lighting installation, compact fluorescent lamps in desk luminaires must provide at least 250 lux on the relevent area of the desk, for which a luminous flux of at least 600 lumen is required.

Glare

Limitation of glare is primarily a matter of proper screening of the lamps by means of suitable luminaires. Research in various parts of the world into the problems of glare has led to the development of different individual national systems for evaluating discomfort glare. Although accompanied by various advantages, the reduction in diameter of fluorescent lamps from 38 mm to 26 mm resulted in an increase in lamp luminance of some 50 per cent, which means that the use of bare fluorescent lamps is even more undesirable than in the past.

Colour appearance

Lamps used in interior lighting are divided into three groups according to their correlated colour temperature (table 1). As can be seen, the limit for group 2 has recently been increased from 5000 K to 5300 K. This is done because it is now agreed that 5000 K lamps belong to the 'intermediate' colour group.

Guidance on where to use the different colour appearance groups is provided by the relevant lighting codes. However, there is a noticeable general trend toward cosier environments, which may gradually lead to a shift of preference from cooler to warmer colours of light.

Table 1 Lamp colour appearance groups

colour appearance group	colour appearance	correlated colour temperature (K) CIE 1975	correlated colour temperature (K) ISO 1981, CIE 1982
1	warm	3300	3300
2	intermediate	3300 - 5000	3300 - 5300
3	cool	5000	5300

Colour rendering

The specification of the General Colour Rendering Index R_a is felt by lighting engineers to be of great help when selecting proper lamp types

Table 2 Steps of minimum values of R_a recommended in different codes

code	recommended steps of minimum values of R_a							
CIE 1975		≥85		≥70	<70			
France 1977	≥90		≥80	≥70	≥60	<60		
Germany 1979		≥85		≥70			≥40	<40
The Netherlands 1981	≥90		≥80			≥50	<50	
ISO 1981			≥80		≥60		≥40	<40
CIE 1982	≥90		≥80		≥60		≥40	<40

Steps of minimum values of R_a recommended in different codes on interior lighting for specific applications are listed in table 2.

The CIE 1975 limit of ≥85 has been changed to ≥80 in the new drafts of CIE and ISO because the R_a of tri-band lamps lies around 85.

In order to eliminate weak points in the R_a concept, certain modifications to the system have been proposed. In addition to the colour rendering index (CRI), a colour discrimination index (CDI) and a colour preference index (CPI) have been suggested.

Contrast rendering

Although contrast rendering in the presence of veiling reflections is primarily a matter of luminaire luminance rather than of lamp luminance, it should be mentioned here that it has been decided to abandon the ESI concept and to adopt the CRF concept internationally for evaluating the effects of veiling reflections. CIE TC-4.1 is busy preparing recommendations for CRF based on a proper reflection standard.

Economy

With continuing increases in electricity prices, cost-effective lighting has become synonymous with energy-effective lighting.

A high luminous efficacy is therefore a basic quality requirement for a lamp. However, lighting designers are fully aware that the efficient use of energy for lighting is not just a matter of applying the most efficient lamp suitable for the purpose, but also of obtaining the highest possible efficiency for the lighting system as a whole, including the influence of the environment. This may be illustrated by the fact that changing over from an unfavourable luminaire to a more favourable one, or from dark room-surface finishes to light ones may yield a more than 20 per cent increase in system effiency.

Physical Chemistry of Light Sources: Certain Principles and Problems.

D. M. Speros, Lighting Research, General Electric Co. Cleveland, Ohio 44112

During the previous Symposium, the observation was made[1] that light source development undergoes evolutionary periods measured in decades or generations. These evolutions occur between revolutionary stages. Based on historical pattern, it was predicted that there should be a revolutionary stage in the non-too-distant future.

The present analysis supports this expectation because it indicates that the critical aspects and limiting problems in lamp science lie currently at the frontiers of existing knowledge. Such a frontier may involve only a number or function, as for example the value of a partial molal free energy and its temperature dependence, or an entire field, such as an unsolved problem in magnetohydrodynamics.

This conclusion is supported by the following corollaries which in turn will be documented by specific examples involving current means of light generation*:

1. A contribution to lamp practice is often a contribution to basic science.
2. The frontiers are extensively interdisciplinary; for example a contribution to lamp practice is usually equally important to practice in seemingly unrelated fields.
3. Simple theoretical considerations or lamp experiments often lead to important conclusions which transcend or overstep an existing frontier.
4. There are frequent long time lags between understanding or basic discovery and application; and vice versa. So it does happen that the reasonable thing is done after all other alternatives have been exhausted.
5. A unifying view or theory of light sources, such as might emerge for example through a concise application of the second law of thermodynamics, is still lacking. Such an overview may well lead to the next revolutionary stage.

Since the emphasis here is in chemistry, this analysis considers first means of chemical generation of light such as flashlamps.

The primer or igniter function of flashlamps relies on the rapid (msec) combustion reaction of solid fuels such as Zr by one or more solid oxidants e.g. $KClO_4$. Solid oxidant - fuel systems are widely applied also in metallothermic processes, propellants and other fields. The problems of modeling and measurement

* See for example Reviews and Bibliographies in Special Issue IEE Proc. 127 (1980). The specific examples given here are taken from work with which the author is the most familiar.

in the simplest case of solid combustion, that of a single particle in oxygen, have proven to be severe. These problems are compounded when large numbers of fuel particles aredispersed among particles of a single solid oxidant and seem to become intractable when a blend of several oxidants is involved. These difficulties were circumvented by the working hypothesis[2] that the combustion of these multicomponent systems is governed by thermodynamically predictable oxidant dissociation pressures and measurable kinetics of oxygen evolution. Selection criteria and a model based on this hypothesis proved consistent with experimental results. Unexpectedly, however, this model predicted that systems thus formulated should necessarily exhibit amplification of the combustion rate i.e. exponential increase in the speed of combustion, if the oxidants were selected so as to dissociate sequentially with increasing temperature. The net effect is that of a "kinetic step ladder" as seen in the Arrhenius plots (solid lines) of Fig. 1 for sequential oxidants Co_3O_4, $BaCrO_4$ and Fe_2O_3. This effect was experimentally confirmed and was applied in the FlipFlash® primer composition. The kinetic ladder principle represents a long time lag between application and understanding because sequential oxidant and fuel formulations have been used empirically for decades. For example, the well known metallothermic "Thermite" reaction involves an oxidant kinetic ladder of the two stage dissociation of Fe_2O_3 at 1300°K and 1800°K and a fuel ladder of Mg and Al.

Chemical flashlamps are being superseded by electronic flashtubes which have also existed for decades in both principle and practice. The current problem of evolution for flashtubes is the same as for other discharge lamps: smaller size and lower cost must be achieved. One of the limiting factors is the cathode which must emit large currents while essentially at room temperature. Since the work function is the energy required to extract electrons by any means, it must be as low as possible. Emitters[3] such as $BaO \cdot Al_2O_3$ with work function ≈1.6 eV, which, for example, helped make possible the use of flashtubes in aircraft, are no longer adequate. Cs compounds with work function less than 1 eV are necessary. The volatility of Cs and Cs_2O is reduced by combining with a more refractory oxide such as MoO_3 to form the molydbate $Cs_2O \cdot MoO_3$[4].

We have established another important requirement: the arc strikes preferentially at the interfacial areas of contact between emitter and metallic substrate because these areas combine the low work function of the emitter with the high electrical conductivity of the substrate. Therefore we developed[5] cathodes which have enormous interfacial areas, the "extended interface porous substrate cathodes," Fig. 2, which consists essentially of a metallic sponge containing the emitter. The combination of Cs compound emitters and enormous

As the halide lamps became possible by the ready availability of vitreous silica, the high pressure Na, HPS, lamps became possible by the advent of dense alumina ceramics. The expectation was that Na would not react with Al_2O_3. In view of so many HPS lamps performing so well for so long it is rather surprising that they are thermodynamically and kinetically unstable over extended ranges of temperature and Na pressure. Over these ranges reactions such as the following[6] occur:

$$Na(l,g) + 2/3\ Al_2O_3(s) \rightleftharpoons 1/2\ Na_2O\cdot Al_2O_3(s) + 1/3\ A\ell(\ell)$$
$$Na_2O\cdot Al_2O_3(s) + 10Al_2O_3 \longrightarrow \beta - Al_2O_3(s)$$

A summary of theoretical and experimental investigations will be given confirming these processes, establishing the ranges of their occurrence and explaining special features, as for example why the blackening caused by the elemental Al (first equation) occurs primarily at the cooler ends where it is often misinterpreted as being due to the cathode. This chemistry is also of importance to fuel cell and electric battery technology where $\beta - Al_2O_3$ is used extensively. Finally, according to the published literature, the evolutionary challenge of using materials other than Na in these alumina envelopes has been met at least to the prototype stage.

Similarities in the chemistry of halogen-incandescent and halide discharge lamps suggest that a unified view might be emerging. For example in both chlorine-incandescent and tin chloride lamps there is a vigorous oxyhalide cycle transporting W up the T gradient, and a cycle involving the disproportionation $2WCl_2 \rightleftharpoons WCl_4 + W$ transporting W down the T gradient.[7] The oxyhalide cycle is responsible for both wall clean-up, and the downfall of both types of lamps: lead attack. The tungsten chloride cycle operating in the oposite direction could replenish the W in the leads, but it has not been possible to-date to balance the two cycles spatially and kinetically in order to avoid both wall darkening and lead attack. Instead, protective coatings have been extensively investigated, especially in fluorine containing lamps, but this, rather than simplifying, complicates the system by indicating that a third cycle is necessary to replenish these cotings continuously.

Analysis of complex, concurrent cycles or mechanisms of chemical reactivity in these lamps seems barely possible without drastic approximations, and experimental confirmation often necessitates new experimental techniques. As an example of approximations due to both omission and commission in such analytical work, one of the mechanisms that has been proposed by the author[8] will be considered. Notable are the omissions of the role of complex halides and also that of oxyhalides.[9] The former reflects the difficulties introduced by data uncertainty (for example, partial molal free energies in the liquid reservoir), by the dissociation behavior of these complexes and by their re-formation thus completing another as yet uncharted cycle. The oxyhalide formation might raise the question of whether, regardless of the stability of Sc, Th, Sn, etc. oxides, W oxyhalides might still form for at least partial wall clean-up. This would involve the implication of off-equilibrium reactions since the oxyhalides of W are volatile while those of Sc, Th, Sn, etc. might not be so. The main omission however is the absence of systematic kinetic studies. This may be thought of as "the kinetic dilemma" because kinetic studies of even a single reaction can be notoriously time - and resource - consuming. A well known example is the reduction of ammonium paratungstate to metallic W[10] which has occupied generations of investigators in universities, industrial laboratories and factories. On the other hand one might reflect on what happened when and since Langmuir studied the kinetics of gas adsorption on lamp filaments.

The main approximation by commission is that local thermodynamic equilibrium, LTE, has been assumed to describe the atmosphere. Actually LTE considerations can be used to show that LTE is not applicable. A simple two cell model, Fig. 3, is used to demonstrate the directions of departure form LTE such as segregation effects, changes in species distribution causing changes in temperature profiles which in turn cause further changes in species distributions and so on, as the system attempts to attain a steady state.

The same model is used to chart the paths of heat content and entropy in an attempt to explore, in terms of the second law of thermodynamics, systems such as this which encompass both randomized and directed motions (heat and work fluxes) in addition to simultaneously occuring radiative emission processes involving threshold energies (quantum systems). Earlier studies[11] of systems proceeding in the opposite direction along similar paths, such as in electroluminescent cells and light emitting diodes, will be reviewed briefly.

A great deal of lamp chemistry is the chemistry of thermal dissociation of solids. Our studies have included $CaCO_3$ (cathodes, phosphors), $Mg(OH)_2$ (ceramics, selective incandescent radiators), Co_3O_4 (oxidants, metallurgical components), $CaHPO_4$ (phosphors), $Cu(OH)_2$ (leads, printed circuits), Fe_2O_3 and $BaCrO_4$ (flashlamps). In order to measure both the thermodynamic quantities such as the integral heat content change, ΔH, and kinetic quantities such as the rate of heat content change, dH/dt, it was necessary to develop a new instrument.[12] Today this instrument is known as the Differential Scanning Calorimeter, DSC, whose use is widespread. A fundamental basis was provided for the empirical equation $da/dt = k(1-a)^m$, (where a is the fraction reacted and $k = A \exp(E/RT)$), which was found by means of these measurements to describe accurately the thermal dissociation of all these diverse solids.

The foregoing examples support the corollaries given at the beginning of this paper. However, it is interesting to conclude with an example related to the additional observation[1], also made during the previous symposium, that the sequence of evolutionary and revolutionary stages reveals the importance of empirical research because theoretical analysis seldom predicts new light sources.

Fig. 4, shows the effects of an axial magnetic field superimposed on an alternating polarity pulsed arc, at 120 pulses/sec, in approximately one atm. of Xe and electrode gap of a few mm. The pulse peak current passing through the coil electrodes in series with the arc is of the oder of 100 amps; therefore the pulsed magnetic field should be of the order of a few hundred gauss. Neither proponent nor opponent magnetohydrodynamic theoretical analyses predicted or explained the behavior seen in the photographs. In retrospect however it can be seen that the magnetic field, rather than causing instabilities, stabilizes the arc (compare the first photograph where the coils have been shorted with the other two photographs with unshorted coils). This stabilization led to applications in precise optical systems. At 75 amps peak current the arc appears constricted in the middle and spreading axially along the coils possibly because of magnetically induced radial pressure. At higher peak current, 120 amp., observation with a rotating sector stroboscope revealed that the arc "explodes" outward in the middle of the gap, after peak current, giving the time integrated appearance of the diamond shape seen in the last photograph.

Even more interesting is the fact that this work was done in 1955 before the efforts on magnetohydrodynamic containment of plasmas were first declassified. So unknowingly and perhaps naively, these largely empirical, and by comparison microscale experiments stumbled into what has since proven to be one of the main frontiers of our age.

References

1. J. F. Waymouth, Proceedings 2nd Intl. Symp. Incoherent Light Sources (1979).
2. D. M. Speros and J. R. Debesis, J. Combustion and Flame 45, 235 (1982).
3. P. B. Davis, et al. U.S. Patent 2,957,231, Oct. 25, 1960.
4. C. I. Cho, et al. U.S. Patent 4,275,330, June 23, 1981.
5. A. Mishra, et al. Patent Application, June 19, 1981.
6. R. E. Hanneman, et al. U.S. Patent 3,453,477, July 1, 1969.
7. D. M. Speros, R. M. Caldwell and W. E. Smyser, High Temp. Sci. 4, 99 (1972).
8. D. M. Speros, R. K. Datta and D. A. Lynch, J. Electrochem. Soc. 128, 778 (1981).
9. See review and bibliography by D. E. Work, Lightng. Res. and Tech. 13, 143 (1981).
10. For example A. B. Kiss, P. Gado and A. J. Hegedus, Acta Chim. Scient. Hungarica 72, 371 (1972).
11. M. A. Weinstein, Phys. Review 119, 499 (1960).
12. D. M. Speros and R. L. Woodhouse, Nature 197, 1261 (1963).

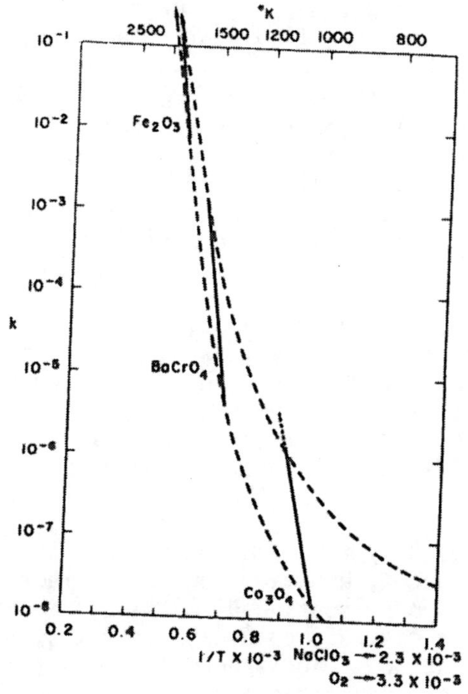

Fig. 1

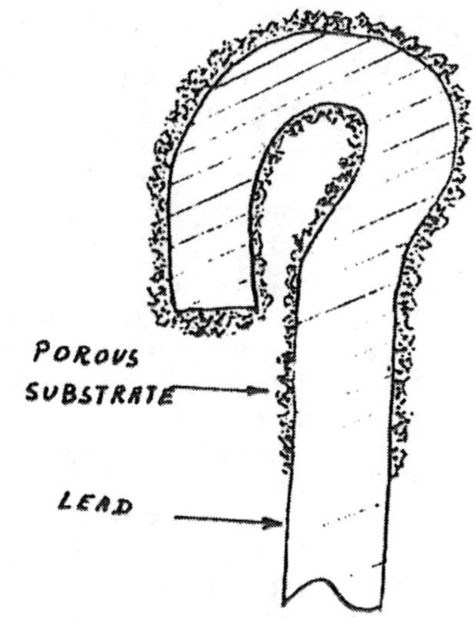

Fig. 2

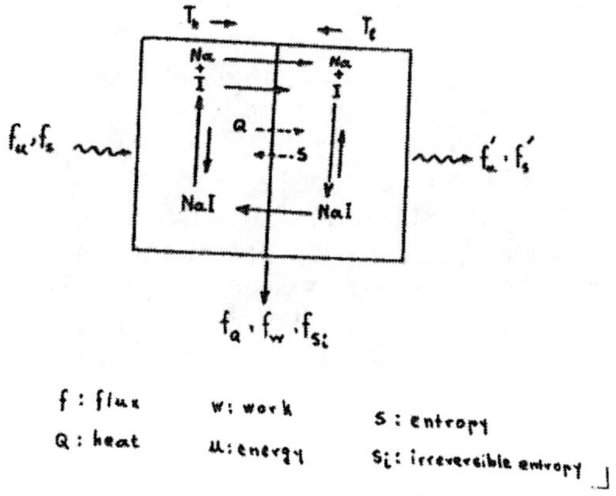

Fig. 3

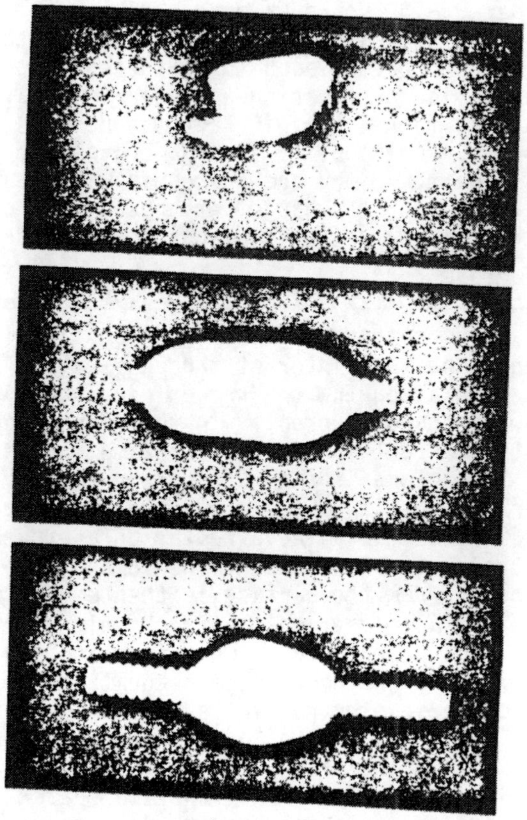

Fig. 4

ELECTRODES FOR ELECTRIC DISCHARGE LAMPS by John F. Waymouth, GTE Products Corp. Sylvania Lighting Center, Danvers, Mass. USA

I. INTRODUCTION

All discharge lamps operated on direct current or alternating current at frequencies up to 100 khz require electrodes for the (cathode) injection of electrons to the discharge and the extraction of electrons from the discharge (anode). Lamps operated on alternating current usually require two identical electrodes because each will serve alternately as anode and cathode. Electron emission being the more demanding of the two requirements the electrode must fulfill, the satisfaction of this function usually is the principal determinant of electrode design, and consequently will be the primary subject of this discussion.

In the main body of the discharge plasma of discharge lamps, the current is carried almost entirely by electrons, since electron mobility is many times greater than that of positive ions. At the cathode surface, however, the ion component of the discharge current is substantially higher. In so-called hot-cathode lamps, in which thermal emission of electrons from the cathode is a major process, the ion current fraction ranges from 5 to 30%. In cold-cathode lamps, in which "secondary" emission of electrons as a result of ion neutralization is the dominant process, the ion current fraction approaches 95%.

The cathode fall of potential, and consequently the "end losses" associated with the discharge lamp, are primarily determined by how many ions each electron escaping the cathode must produce in order to insure continued emission of electron current, and by the ionization potential of the gas. In cold-cathode lamps, each electron must typically be responsible for creating by ionizing collisions ten-to-twenty ions to return to the cathode, because the coefficient of secondary electron emission is only five-to-ten percent. Consequently, cathode fall is high, of the order of 100 volts or more.

In hot-cathode lamps, because of the greater efficiency of the electron emission process, each three-to-ten electrons escaping the cathode need only be responsible for creating a single ion that returns to the cathode. Therefore, cathode fall is much lower, in fact may be less than the ionization potential of the gas, when current density is high enough for two-step ionization to occur. Because power dissipated in the end losses is wasted so far as the production of light is concerned, most discharge lamps used in general illuminating service employ hot cathodes. For that reason, this discussion will deal entirely with cathode phenomena in hot-cathode lamps.

II. THE CATHODE-TO-DISCHARGE JUNCTION: A PROBLEM OF INTERDEPENDENT ZONES

The cathode-to-discharge junction is a transition in which widely different phenomena must interact through several different "zones" constrained always by the consideration that the sum of electron current plus ion current is always equal to the discharge current. In this discussion, I will deal only with the DC case of total current independent of time, which is in itself quite complex.

A. Zone 1: The Cathode Surface The most important zone is the cathode surface itself, which may be characterized by a work function Ø, determined by the additional energy required by an average electron to escape from the solid. The cathode surface is maintained at some temperature T. The electron emission current density from the surface is then determined by the work function, the temperature and the accelerating electric field at the cathode surface.

The dependence on work function and temperature is, of course, that of the familiar Richardson-Dushman equation: J = A(T**2)exp(-eØ/kT), where Ø is the work function in electron volts, and A is a material-dependent constant.

The influence of accelerating electric field enters in four ways, at progressively increasing field strengths: first, overcoming surface retarding fields due to non-uniform work function (Patch Effect, 0-10^4V/cm); second, reducing the apparent work function barrier due to image-charge effects (Schottky Effect, 10^4-10^5V/cm); third, reducing the thickness of the work function barrier to the extent that energetic electrons which do not have sufficient energy to surmount the barrier may nevertheless tunnel through it (temperature-field emission, 10^5-10^7V/cm); and, finally, reducing the thickness of the work function barrier to the extent that even electrons of zero-point energy can tunnel through it (Field emission, 10^7-10^8 V/cm). In the first three modes of interaction of electric field, electron emission current density is a function of both temperature and electric field. For field emission, electron emission current density is substantially independent of cathode temperature.

In the Patch Effect range, overcoming the retarding field exerted by adjacent high work function patches on electrons attempting to escape from low work function patches results in an increase of emission current density by factors of ten to fifty between zero accelerating field and the point of saturation (a few thousand volts/cm). In the Schottky effect range, increases in thermionic emission owing to a lowering of the work function barrier amount to factors of 2 to 10 between acceler- ating fields of 10^4 and 10^5 volts/cm. In the temperature-field emission range, e- lectron emission current densities increase by factors of 100 to 1000 between ac- celerating fields of 10^5 and 10^7V/cm. In the field emission current range, emission current densities increase by factors of 100 to 10000 over the field strength range of 10^7 to 10^8V/cm. Overall, therefore, between zero field and 10^8V/cm accelerating field at the cathode surface, the electron emission current density may increase by as much as a factor of 10^{10}.

B. Zone 2: The Space-Charge Sheath The extreme sensitivity of cathode e- mission current to accelerating electric field at the cathode surface necessarily implies a very strong leverage for the factors which determine the accelerating electric field at the cathode surface. The principal such factor is the ion current density reaching the cathode. This dependence arises from the fact that when ion current amounts to more than a few percent of the electron current, the ion number density greatly exceeds the electron number density, because the heavy ions move much more slowly than the electrons. Consequently, immediately in front of the cathode, there exists a space charge sheath in which there is a net positive charge. To the approximation that electron density may be neglected in comparison with ion density, and that this sheath is thin in comparison with a mean-free-path for col- lision, ion and electron motion are free-fall ("inertial") and the electric field at the cathode surface can be calculated from the Langmuir-Child Law: $\mathcal{E}_k = 2(J_+/\epsilon_0(M_i/2e)^{\frac{1}{2}})^{\frac{1}{2}}V_k^{\frac{1}{4}}$ Volt/meter, where J_+ is ion current density in amp/meter^2; M_i is ion mass in kilograms; ϵ_0 is the permittivity of free space, $10^{-9}/36\pi$; e is the e- lectron charge, 1.6×10^{-19} coulomb; and V_k is the potential across the space charge sheath, usually a major component of the cathode fall.

C. Zone 3: The Ionization Region or Negative Glow In Zone 3, the plasma region outside the space charge sheath, are created the ions which traverse the sheath to create the electric field at the cathode surface. These are created by collisions with atoms of the plasma gas by electrons which have acquired kinetic energy in traversing the space charge sheath. The ion current density reaching the sheath boundary is then a complex function of the electron current density incident on the ionization region, the value of the sheath potential, the nature and pressure of the plasma gas, and the presence of electric fields within the ionization region. the ion-to-electron-current ratio delivered by this zone to the sheath is in general extremely difficult to calculate. In low-pressuredischarges (fluorescent lamp), it is best measured experimentally. In model calculations for the high pressure dis- charge (HID) case, it can only be dealt with by relatively crude approximations.

D. Zone 4: The Cathode Structure The cathode structure (geometry and ma- terials) serves to establish the energy dissipation required to obtain a desired temperature for the cathode surface; it may also be a reservoir of activating ma- terial for the purpose of maintaining a low-work-function emitting surface through- out life.

III. SOLUTION

The interdependence of the four zones can best be expressed in terms of a coupled system of simultaneous equations: 1) $I = I_+ + I_-$; 2) $I = \int JdA$;

3) $J_- = f_1(T_k, \emptyset, E_k)$; 4) $E_k = f_2(J_+, V_k)$; 5) $J_+ = f_3(V_k, J_-)$;
6) $T_k = f_4(P_k, I, \emptyset, \text{geometry})$; 7) $P_k = I_+V_k + I^2 R_k + f_5(I, V_k)$ Equation 1) is the equation of continuity of total current; equations 3) and 4) determine electron emission current density as a function of cathode temperature and ion current density. Equation 5) determines ion current density from the ionization region. Equations 6) and 7) describe the cathode power balance, determining its temperature from a), power delivered to it by ion bombardment and other processes, and b) its power dissipative capability.

IV. APPLICATION TO LAMPS

A. Fluorescent Lamps In the case of fluorescent lamps, it proves to be possible to measure experimentally the value of many of the important intermediate variables, in particular the ion current delivered by the ionization region to the sheath for a particular case, as well as the zero field thermionic emission. The models for the remainder of the equations are adequate to permit calculation of all the rest of the variables; this is discussed in detail in reference 1. From these results, it can be concluded thatthe fluorescent lamp with oxide cathode is a "patchy" emitter whose external surface consists of patches whose work function differ by 1-1.5 electron volts; the "pores" between particles are identified as the low work function patches, and the exposed crystal faces are the high work function faces. Ion current densitities of 50 ma/cm^2, producing average surface electric fields of $\sim$ 8500V/cm, increase electron emission tenfold above the typical zero field emission of 100-200 ma/cm^2.

B. H.I.D. Lamps The case of HID lamps is somewhat more difficult, because none of the variables except average electrode tip temperature can be measured. An attempt to analyze this case has recently been carried out by Waymouth.[2] It is found that there are usually two solutions to the system of equations. One is a low current density, low electric field mode in which electron emission is Shottky-amplified thermionic emission. This is the so-called "diffuse" mode, commonly found in high pressure sodium lamps, where the arc footpoint covers most of the end of the cathode. The other is a constricted or "hot-spot" mode, frequently found in metal halide lamps, in which the emission mechanism is temperature-field or field emission, and the arc footpoint is much smaller than any dimension of the cathode; current densities typically exceed 10^5 amp/cm^2. The mode which is favored in any given case is the one which has the lowest cathode fall and the lowest cathode operating temperature. The hot-spot mode is favored by high operating pressure, and, surprisingly, by low work function. The preference for the hotspot mode in the case of low work function results from the fact that field emission current density varies as exp $(-\emptyset^{2/3})$, while thermionic emission varies as exp $(-\emptyset)$. Thus, although thermionic emission current density is increased by decreasing work function, field emission current density is increased even more.

(1) J. F. Waymouth, Electric Discharge Lamps, M.I.T. Press, Cambridge, Mass. 1971 Chapter IV.
(2) J. F. Waymouth, Journal of Light and Visual Environment, Vol. 6, No. 2, winter 1982.

SUMMARIES OF PAPERS

ORAL AND POSTERS CONTRIBUTED PAPERS

II

COMMUNICATIONS ORALES ET AFFICHEES

RESUMES DES COMMUNICATIONS

n° 1

The SEF Lamp; A Compact Energy-Efficient Replacement for the Incandescent Lamp

John M. Anderson
General Electric Research & Development Center
Schenectady, New York, USA

Fluorescent lamp technology can give high efficacy and long life, but electrode losses are emphasized and efficacy suffers if the lamp is simply shortened to the size of the incandescent lamp in an attempt to devise a compact light source. Folding the lamp retains the ratio of column to end losses, but complete elimination of the electrodes by excitation of the plasma with a solenoidal electric field (SEF) can give a quite small toroidal plasma which efficaciously excites the phosphor. A ferrite toroid core is chosen to give tight coupling through transformer action between the electronic ballast generating 100 kHz power and the plasma as a one-turn secondary.

A 2000 lumen lamp, 7.6 cm in diameter and 15 cm overall, with integral ballast can have more than 50 lumens per watt. Frequency of operation with presently available ferrites is 100 kHz. The ferrite is supported in the discharge space and is excited by a ten-turn primary winding, the wire of which is protected by thin glass sleeving.

Narrow-line phosphors, with red and green emission, as developed by Philips Co., cover the ferrite, and the inside surface of the glass envelope. The blue emission from mercury complements the phosphor to match closely the color temperature of incandescent. The ferrite core temperature was measured at 300°C (Curie temperature ∿370°C) and envelope temperature at 100°C for a lamp with 30-35 watts to the primary winding. Thus mercury pressure needs control and this was done by a Pb/Bi/Sn alloy, also described by Philips workers.

Krypton was found to be a better choice for fill gas than argon for two reasons, its lower discharge voltage reduced ferrite core loss to ∿3 watts, and its inherent "positive column" efficacy was higher at low gas pressures and high loading. We measured the positive column efficacy for a freshly-constructed lamp at about 70 lumens per watt for a rare gas pressure of 0.5 Torr. Considering ferrite core loss, light loss by attachment of lamp to ballast shell, and ballast efficiency, say 90%, an overall lamp efficacy greater than 50 lumens per watt is attainable. The lamp is dimmable, has instant start and restart and can be made in a lumens package from ∿500 to 2500 lumens.

n° 2

"PL" LAMP WITH AN INTEGRAL STARTING DEVICE

W. Veenstra

Nederlandse Philips Bedrijven B.V., Lighting Division

Roosendaal

The Netherlands

Recently two new families of compact energy saving fluorescent lamps have been introduced. They are the "SL" lamps with an integral starter and ballast and the "PL" lamps in which only the starter is integrated with the discharge tube. Both families of lamps are based on the use of narrow diameter fluorescent tubes.

The properties of the "PL" lamps, including the measures taken to control the mercury vapour pressure, have been discussed in earlier papers [1,2). In this paper special attention will be given to the combination of the discharge tube and the glow switch starter which forms the "PL" lamp.

The general requirements for the discharge tube - glow switch starter combination are as follows:

- ignition time shorter than 2 seconds
- no influence on the lamp performance due to the glow switch starter
- good life time

With the narrow diameter discharge tube used in "SL" and "PL" lamps, there is a relatively high reignition voltage after each reversal of the discharge current. This is an additional complication in "SL" and "PL" lamps compared with fluorescent lamps with wider diameter tubes where the reignition voltage is less. For the "PL" lamp there is the added complication that the same lamp has to function correctly on both 118V and 220/240V supply voltages.

A study has been made of the parameters which influence the reignition voltage of the discharge tube, e.g. tube length, mercury vapour pressure and gas impurity levels. Thus it was possible to determine the situations for which the highest reignition voltage will be found.

The relatively high reignition voltage associated with the narrow diameter discharge tubes also appears across the glow switch and causes a glow current to flow through this element. If this glow current exceeds a certain value it will cause the switch to close resulting in a lamp which repeatedly extinguishes and reignites. Thus a glow switch with a low glow current is required. However a glow switch with a short ignition time is also required and these two requirements are contradictory. This led to the development of a special glow switch where the correct balance between these two aspects was found.

Combining the discharge tube and glow switch data, it is possible to develop a "PL" lamp where the reignition time is less than 2 seconds, where the lamp performance is not influenced by the glow switch and where the life time is in excess of 5000 hours.

References:

1. A. Bouwknegt, J. Illum. Eng. Soc. 11, 204, (1982)
2. L.E. Vrenken and W. Veenstra, paper presented at CIBS National Lighting Conference, Warwick, U.K. (1982)

n° 3

COMPACT FLUORESCENT LAMP

by

J. M. Proud

GTE Laboratories, Inc.

40 Sylvan Road, Waltham, MA 02254

A compact fluorescent lamp has been the target of many research and development efforts in recent years in the quest for a cost-effective, efficient alternative to the incandescent lamp. Fluorescent devices offer the features of efficiency, longevity, and instant response which are essential properties of an incandescent replacement product. However, the long positive column discharge and negative V-I characteristics, intrinsic properties of the classical fluorescent lamp, have been impediments in developing compact light sources from this technology.

The abandonment of the positive column discharge in favor of electrodeless, radio frequency excitation has yielded encouraging results in terms of the uncomplicated, compact shape of the discharge vessels and the reasonable efficacy.(1,2) Most of this work has been accomplished with radio frequencies below 50 MHz, a choice probably influenced by the existence of ISM bands at 13.56, 27.12, and 40.68 MHz.

This paper describes a compact fluorescent lamp which is excited at 915 MHz.(3,4) Analysis and experiment demonstrate the advantages at this frequency in producing a highly uniform, efficacious discharge with capacitive coupling. The plasma impedance in such a discharge tends toward a value which matches the source impedance with electron density in the range, 10^{11} to 10^{12} $cm.^{-3}$ No tuning or matching devices are required, and the discharge uniformity permits optimization at efficacy levels above those achievable at lower frequency where inductive coupling prevails.

A compact fluorescent lamp is described with variable lumen output in the 500 to 3000 lm range. Efficacy near 100 lm per radio frequency watt are achieved at 15 RF W. Lumen maintenance is approximately 95% at 1000 h with 20 W input. Envelope temperatures below 60°C are observed for the lamp which resembles an incandescent lamp in both size and shape. The lamp is self-starting with the application of normal operating power.

References

1. D. D. Hollister, U.S. Patent No. 4,010,400, March 1977.
2. H. H. Glasscock, U.S. Patent No. 4,005,330, January 1977.
3. J. M. Proud and R. K. Smith, U.S. Patent No. 4,266,166, May 1981.
4. J. M. Proud and D. H. Baird, U.S. Patent No. 4,266,167, May 1981.

n° 4

POSITIVE COLUMN CHARACTERISTICS OF SMALL DIAMETER FLUORESCENT LAMPS

S. L. ZHU and B. H. ZHANG
Institute for Electric Light Sources, Fudan University, Shanghai, China.

It is well known that the fluorescent lamps have much higher luminous efficiency than that of conventional incandescent lamps. However, the use of the fluorescent lamps is limited because of their large tubular sizes. Recently, with the development of the three-band phosphors, which have relatively long life at high loading, the small diameter fluorescent lamps with internal diameters from 9 to 14 mm aroused the interest of lamp engineers, since they possess the characteristics of compactness and can further be compacted by folding. It is thought that they might replace the incandescent lamps in most cases. In short, they have great potential as energy saving lamps. To our knowledge, although the conventional fluorescent lamps are well understood(1), little work has been done on the small diameter fluorescent lamps. It is the object of this paper to study the column characteristics of these lamps.

The experimental discharge tubes were those with the diameter of 14 mm and the rare gas filling at 3 torr. Langmuir probe techniques were employed to measure the electron temperatures and electron densities at discharge currents from 50 ma to 500 ma and wall temperatures from 24°C to 60°C. A floating probe technique was used to measure the axial electric fiels at the same conditions. The 253.7 nm radiation power measurements were carried out by a quartz discharge tube with the same size and the gas filling. The results were compared with a simple theory of Lama et al(2). Because of the large losses in small discharge tube, the axial electric fields are higher than that of conventional 38 mm diameter tube. Consequently, the electron mobility is decreased considerably. If the experimental vale of the electron mobility is applied in Lama's theory, there is a reasonable agreement between the theory and the experiments.

It is concluded that after this correction, the Lama's simple algebraic formulas are sufficient to be used to predict the discharge parameters of the small diameter fluorescent lamps.

REFERENCES

(1) J. Waymouth, Electric Discharge Lamps, MIT Press, Cambridge, 1971.

(2) W. L. Lama, C. F. Gallo, J. J. Hammond and P. J. Walsh, Applied Optics, Vol. 21, N°. 10, 1801-11, May 1982.

THE THREE-BULB LAMP

A.L.J. Burgmans and P.R. van IJzendoorn
Philips Research Labs., Eindhoven, The Netherlands

The increasing energy price has stimulated the search for a substitute of the incandescent lamp. An alternative is the compact fluorescent lamp which combines the high efficiency and the long life of the conventional fluorescent lamp with the small size of the incandescent lamp. The recently introduced "SL" lamp is an important step towards this goal. Although its dimensions are still larger than those of an incandescent lamp, the "SL" is provided with a screw socket and thus exchangeable with an incandescent lamp. The system efficiency of the "SL" is four times higher than that of an incandescent lamp. Efforts to make even more compact lamps imply that the discharge tube must be folded more than three times as is presently the case. To avoid the difficulty of bending a discharge tube many times, the discharge can also be folded, and in a more compact way by what is referred to as the multi-bulb technique. The two-bulb lamp has become known from Yamazaki [1] and Wesselink's[2] work; the lamp consists of two telescoped bulbs. The inner bulb is provided with grooves which determine the discharge channel. Compactness is enhanced by increasing the number of grooves. As a consequence, however, the distance between the grooves decreases so that there will be a greater risk of breakdown through the narrow gaps separating the grooves. Hence the required dimensional accuracy of the bulbs increase as the compactness of the lamp increases.

A compact fluorescent lamp based on the multi-bulb concept which requires a less stringent dimensional accuracy than the two-bulb lamp is the three-bulb lamp[3]. This lamp consists of three telescoped bulbs as is shown in the figure. The entire space between the inner and the outer bulb is filled with a discharge. The discharge path between the two electrodes is determined by the intermediate bulb. This path is located alternately at the inner and outer surface of the intermediate bulb owing to holes at the bottom of this bulb. Between these holes the discharge travels through an U-shaped path over the lamp dome. A fluorescent powder is present on those parts of the bulbs that are in direct contact with the discharge. The flat analogue of the above lamp is essentially composed of two flat plates, an upper and a lower plate with an intermediate plate which is located alternately near the upper and lower plate.

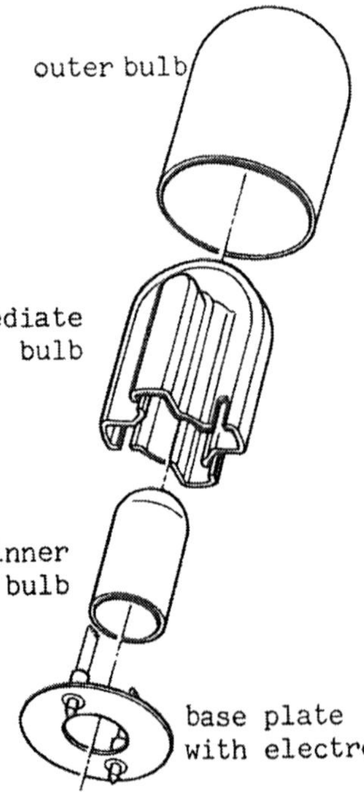

The length of the discharge channel of all lamps that have been made is 40 cm and the cross-sectional area of the discharge is about 1 cm^2. For the pear-shaped lamps the discharge channel has been folded into twelve approximately equal parts within an outer bulb with a diameter of 5.7 cm and a total height of 5 cm. The tile-shaped lamps are circular with a diameter of 10 cm and a height of 1.4 cm, the discharge channel being folded into 7 parts. Apart from the buffer gas the pear-shaped lamps contain an amalgam and the tile-shaped lamps a mercury droplet. With an input power of 15 W both types stabilize at their maximum light output, which is 900 lm for the pear-shaped lamp and 1050 lm for the tile-shaped lamp.

1. H. Yamazaki and H. Akutsu, U.S. Patent No. 4095135, 1978.
2. G.A. Wesselink, Philips J. Res. 35, 337-353, 1980.
3. A.L.J. Burgmans and P.R. van IJzendoorn, Dutch Patent Application No. 8001280.

ANALYTICAL MODEL FOR LOW PRESSURE GAS DISCHARGES: APPLICATION TO THE MERCURY + ARGON DISCHARGE

W.L. Lama, C.F. Gallo*, T.J. Hammond
Xerox Corporation, Rochester, N.Y. 14644
and
P.J. Walsh
Fairleigh-Dickinson University, Teaneck, N.J. 07666
*Present Address: 3M Center, Bldg. 201-1E, P.O. Box 33221, St. Paul, MN 55133

ABSTRACT

A general but simple technique for analyzing complicated gas discharges has been developed and successfully applied to the Hg + Ar discharges that exist in fluorescent lamps. The Hg + Ar theoretical model includes electron excitation, electron de-excitation, two-stage mercury ionization through a saturated metastable level, and a proper treatment of the self-absorption of the Hg 2537Å radiation. Under suitable conditions, the analysis yields simple analytic expressions for the electron temperature, the mercury 2537Å radiation and the electrical field. The electron temperature takes the following functional form, which is applicable to a wide range of discharges.

$$\frac{kT_e}{eV_i} = [\ell n\ (C_o N_A N_o R^2)]^{-1}$$

where N_A is the argon density, N_o the mercury density, R the tube radius, V_i the Hg ionization potential and all C's are calculated constants. The 2537Å power per unit length (P) is given by

$$P = \frac{C_1 (N_o/N_A)^{\frac{1}{2}} N_e R}{1 + C_3 N_e N_o R}$$

where the electric field is given by

$$E = \frac{(C_4/R)\ (N_o N_A)^{\frac{1}{2}}}{1 + C_3 N_e N_o R} + \frac{C_5 N_A^2}{[\ell n\ (C_o R^2 N_A N_o)]^{3/2}} + B$$

where the first term is related to Hg 2537Å radiant energy, the second term is due to elastic scattering loss and B is related to other energy loss terms which are either small and/or comparatively weak functions of the pertinent parameters. When applied to Hg + Ar discharges, these expressions yield good quantitative agreement with the available absolute data on the dependence of the Hg 2537Å radiation and electric field on mercury pressure and current. All of the parameters used in the Hg + Ar calculations are derived from basic data available in the literature. Contrary to some earlier descriptions of these discharges, our treatment requires no adjustable parameters.

n° 7

Experimental Studies of Enhanced Emittance by Zeeman Splitting of a Low-Pressure Mercury/Argon Discharge

by

D. Hollister
and
S.M. Berman
Lawrence Berkeley Laboratory
University of California
Lighting Systems Research

Abstract

The Zeeman-effect experiment is designed to quantify the amount of mercury resonance radiation production at λ2537 at elevated (i.e. >42°C) cold-spot temperatures in a mercury discharge tube in the presence of an external magnetic field.

Theory predicts two effects:

a) Absorption and subsequent loss of resonance radiation have been shown to depend on a critical mercury number density, $n \simeq 2 \times 10^{13} cm^{-1}$ per isotope. The creation of three well-separated magnetic sub-states via the Zeeman effect is expected to increase this critical number density by a factor of three.

b) If the Zeeman displacement is comparable to the Doppler width of the individual lines, the inter-line exchange of photons within overlapping line wings enables the escape of an even further enhanced number of photons. This enhancement has a predicted value of an additional factor of 1.7. Experimental verification of this additional enhancement would constitute an important substantiation of the theory.

The experimental apparatus consists of an electrically shielded pair of short solenoids arranged in a quasi-Helmholtz configuration to provide a relatively uniform axial magnetic field within the observational volume of a mercury discharge tube. Provision is made to regulate the temperature of the mercury cold-spot reservoir between the limits $40°C \leq T \leq 70°C$ with a resolution of 0.1°C.

A 3.4-meter Jarrel-Ash grating spectrograph is employed in the 22nd order (58°) to view the hyperfine structure (hfs) of the resonance line of natural mercury at λ 2537. Data are taken photographically on Kodak Type SA-1 plates at constant input current and power as functions of applied magnetic field and cold-spot temperature.

Preliminary results indicate that the experimental system is conceptually sound, but the resolution of the Type SA-1 plates (≈ 100 lines/mm) has been found insufficient to fully resolve the σ components of the hfs. This problem is expected to be solved by substitution of Kodak Type 649 plates, the resolution of which is ≈ 1500 lines/mm.

If all theoretical predictions hold true, the anticipated result of this study is a shift in the temperature of peak resonance radiation production from 42°C to 63°C, with a concomitant increase in resonance radiation emission of about a factor of five.

n° 8

Experimental Studies of Enhanced Emittance by Isotope Enrichment of a Low-Pressure Mercury/Argon Discharge

by

R.K. Sun
and
S.M. Berman

Lawrence Berkeley Laboratory
University of California
Lighting Systems Research

Abstract

Natural mercury has seven stable isotopes that will, when excited in a low-pressure gas-discharge tube, generate UV lines due to transitions from the 6^3P_1 state to the ground state at slightly different wavelengths. Wishing to investigate the increase of UV radiation (254 nanometers) for different mixtures of Hg isotopes that would lead to improvements in the efficacy of fluorescent lamps, we have prepared an experimental set-up for a specially designed discharge tube that will be used repeatedly to test various isotopes and isotope mixtures. The tube is provided with a heat pump unit to control the cold-spot temperature of the mercury sample and a vacuum pump aggregate to regulate the rare-gas pressure of argon inside the tube. The tube, made of quartz, is enclosed in a plastic box that is flushed continuously with nitrogen or argon gas to prevent the generation of ozone and to absorb UV radiation. The current through the discharge tube is controlled by a power supply that can provide different voltages and loads and can maintain a sinusoidal waveform for its positive column current. The wall of the tube is maintained at a temperature that is constant and higher than the temperature of the mercury sample, which determines the mercury vapor pressure inside the tube.

Theoretical predictions indicate a sizeable enhancement with a shift in temperature of peak resonance radiation production. Quantitative experimental results will provide a test of the theory.

n° 9

DISCHARGE LOW PRESSURE IN MERCURY

APPLICATION TO THE TREATMENT OF WATER

AUTHORS : J. MOURIER AND G. LE MINTIER

Among the different devices of desinfection of water, this of the action of the line 253,7 mm of mercury offers many advantages.
This device is effective for almost all germs and virus without formation of cancerigenics organochlorinated products.

LAMP

The lamps are of the low pressure type in a discharge of mercury. Taking into account their operating conditions in a quartz sleeve, which insulates them from the liquid to treat, they are provided with two cavities at the ends.

Their electric power (26 or 56 W) delivers for the line 253,7 mm a contact current respectively for 37 000 or 47 000 W/cm^2. In these conditions they are efficiency for debits being able to go to 3 000 m^3/h in potability and 300 l/h in sterilization.

As every discharge lamp they are stabilized by a ballast, and their ignition is secured by a glimmer starter identical with this of fluorescent lamps.

SPECTRUM

The spectrum radiated is this of low pressure mercury ; the line 253,7 mm represents more than 50 % of mains operated electric power.

IRRADIATOR

The lamps work in installations specially conceived to obtain an efficiency destruction of germs and virus.
- in glass, we can easily control the good working of the lamp, the quality of the water and the dirty mark of the sleeve. A photo-electric cell, put outside of the irradiator, points out either visually (amperemeter lamp) or in this sound way, every anomaly.
- in stainless, they are provided with an ultrasonic diaphragm which prevents the deposits on the quartz sleeve.

APPLICATIONS AND RESULTS

The irradiators have been tested with success for the following situations :
- sterilization for the use of hospitals and pharmacy ;
- treatment for the home use - WATER SUPPLY OF PARIS ;
- aquariums of NANCY - MONACO ;
- aquiculture in fresh or salt water.

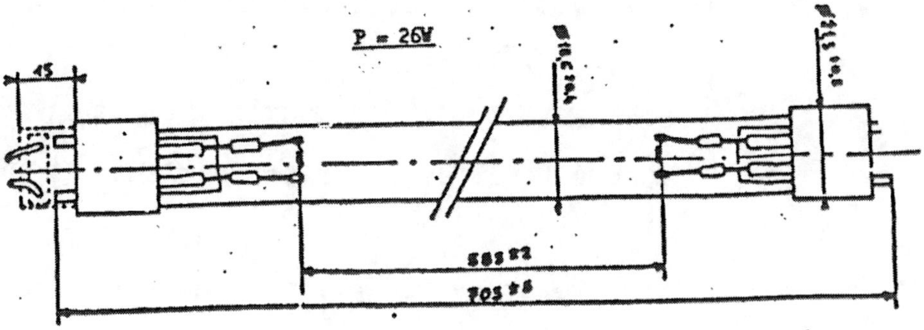
P = 26W
15
585 ±2
705 ±5

LAMP UV B.P.

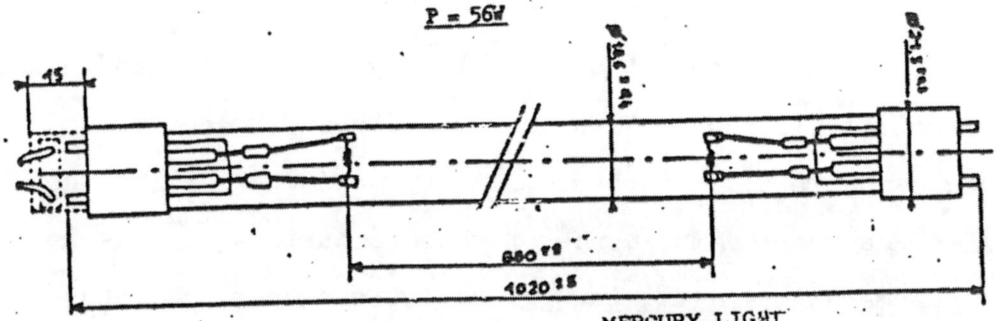
P = 56W
15
880 ±2
1020 ±5

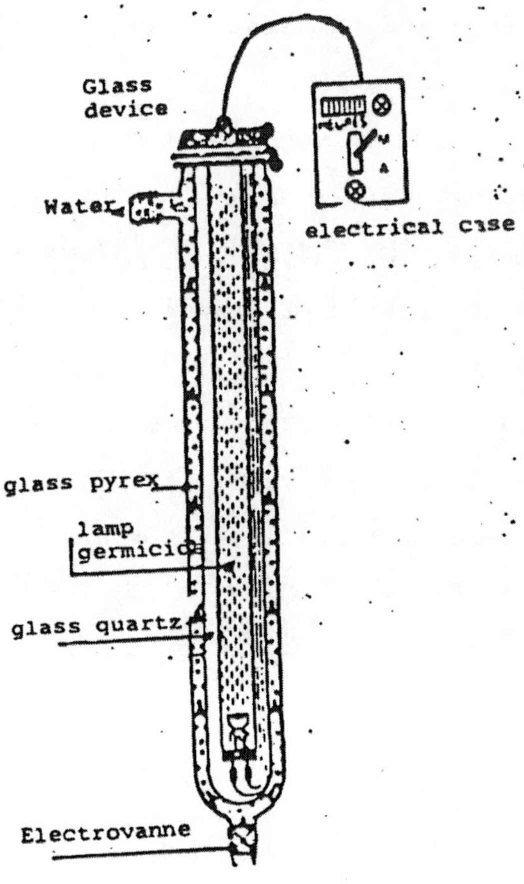
Glass device
Water
electrical case
glass pyrex
lamp germicid
glass quartz
Electrovanne

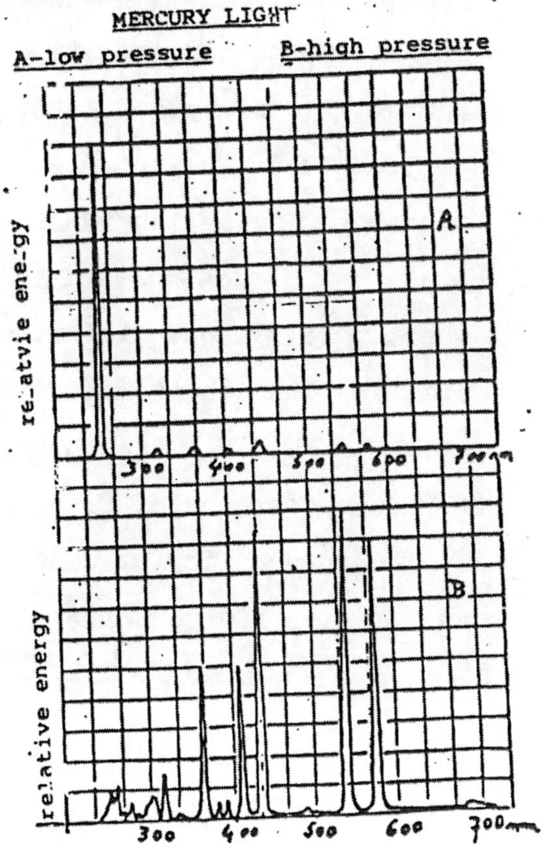
MERCURY LIGHT
A-low pressure
B-high pressure
A
B
relative energy
300
400
500
600
700nm

n° 10

A STUDY OF THE HOLLOW CATHODES FOR MERCURE DISCHARGE LAMPS

S.P.RESHONOV, N.F.ANTOSHKIN, S.L.RIBALOV
Moscov Power engineering institute (USSR)

1.Introduction. We study the plasma generation in the active zone of hollow cathode (HC) for low and high pressure mercury lamp (Hg+Ar), by thermoionic emitted electrons whose values of energy are close to e exitation or ionization energy of gas. Elastic and inelastic electron-atom collisions and coulomb collisions between emitted and plasma electrons are taken into account. By means of solving the closed system we analyse the influence of the arc current (I), geometrical cathode parameters, and filling gas conditions upon the cathode drop (U_c), plasma electrons concentration (n_e) and temperature (T_e), part of ionic current by the cathode $\mu(r_c)$, etc.

3. Method. Unlike [1,2], we have solved the kinetic Boltsman equation by using new boundary condition on surface of the cathode:

$$\left[\partial\Psi(\varepsilon,\xi)/\partial\xi + U_+(\varepsilon)\cdot\beta(\varepsilon)\cdot\Psi(\varepsilon,\xi)\right]_{\xi_0} = U_+(\varepsilon)\cdot f(\varepsilon)$$

where $\Psi(\varepsilon,\xi)$-nonequilibrium correction of maxwellian electron energy distribution function (EEDF) [1]; $\varepsilon=(E-eU_c)/T_e$, E is the electron energy; $\eta_1= E_1/T_e$, E_1- the threshold of inelastic processes; $\xi=r/L$, $\xi_0=r_0/L$, r-radial coordinate; $U_+(\varepsilon)=0$, $(\varepsilon\leqslant 0)$; $U_+(\varepsilon)=1$, $(\varepsilon>0)$.

The nonequilibrium correction of maxwellian EEDF have been given by the inverse finite Gankel transformation:

$$\Psi(\varepsilon,\xi) = \beta_0(\varepsilon,\xi) + \frac{2\cdot exp(\varepsilon/2)}{\xi_0^2}\cdot\sum_{m=1}^{\infty}\frac{K_m^2\cdot J_0(K_m\cdot\xi)\cdot\varphi(\varepsilon,K_m)}{[K_m^2+\beta^2(\varepsilon)\cdot U_+(\varepsilon)]\cdot J_0^2(K_m\cdot\xi_0)}$$

where $\beta_0(\varepsilon,\xi)$ is the initial member of the Dini series of $\Psi(\varepsilon,\xi)$ [3]; $k_1, k_2, \ldots$-positive roots, and $\pm ik_0$ -imaginary roots of en equation:

$$K\cdot J_0'(K\cdot\xi_0) + U_+(\varepsilon)\cdot\beta(\varepsilon)\cdot J_0(K\cdot\xi_0) = 0$$

$\varphi(\varepsilon,K)$ is the finite Gankel transformation of $\Psi(\varepsilon,\xi)$;

$\Psi(\varepsilon,\xi)$ can be used for the calculation of inelastic collisions probability and energy transformed by emitted electrons to plasma ones.

3. Result and discussion. Our calculation were carried for the thin-wall (0,1mm) HC of the high pressure mercury lamp with bore diameter (d_c) 0,1-0,2 mm, with argon filling gas (10÷40 mm Hg) and current I=1÷4 A. At low pressure the heavy particles gas keep an equilibrium with the cathode. With the argon or mercury vapour pressure rise, heat exchange between electron and heavy particles increases by elastic collisions and "an alienation" of atom temperature from cathode temperature takes place. As a result the neutral gas begin to take part in the cathode heating. The value $\mu(r_c)$ decreases (fig.). Similar calculations are taken for HC of the fluerescent lamp; optimal value of diameter of HC exists and can be found with the help of this method.

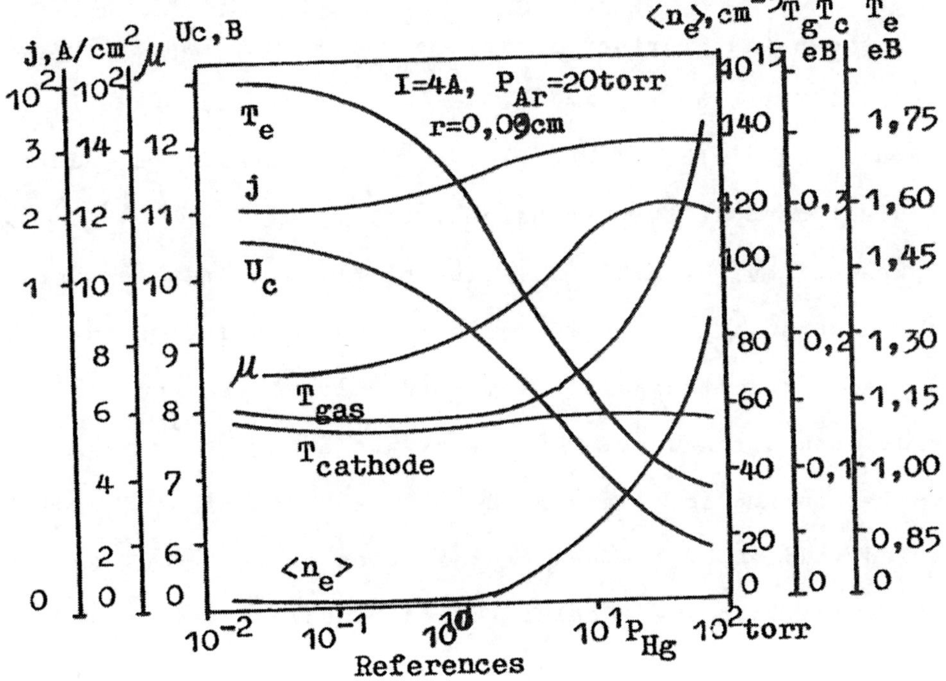

References

1. S.P.Reshonov, A.Lunk, Beitr. aus der Plasmaphysic, *18, (1978), 101; 381.

2. S.P.Reshonov, N.F.Antoshkin, Proc. 15-th Inter. conf. on phen. in ionized gas. Minsk, 1981, p.2., 757.

3. G.N.Watson. A treatise on the theory of Bassel functions, 1945.

LOW PRESSURE SODIUM LAMPS WITH A LUMINOUS EFFICACY OF 200 Lm/W

L. Sprengers, R. Campbell* and H Köstlin°
N.V. Philips Industrie, Turnhout, Belgium
*North American Philips Lighting Corporation,
Hightstown, New Jersey, U.S.A.
°Philips GmbH Forschungslaboratorium Aachen, Germany

During the last 50 years the low pressure sodium lamp has undergone a series of improvements which have increased the luminous efficacy from about 38 lm/W to a value in excess of 180 lm/W. From a study of the characteristics of the low pressure sodium discharge and the power balance of the present day low pressure sodium lamp, it is shown that further improvements in the lamp efficacy are still possible. This can be realized by a further improvement in the thermal insulation of the discharge tube, together with a reduction in the power dissipated in the discharge tube.

The present day thermal insulation of the SOX lamp consists of placing the discharge tube in an evacuated outer bulb. On the inside of this bulb there is a transparent semiconducting layer consisting of indium oxide with a tin dope. This layer has a high transmission for visible radiation and is a very good reflector of infrared radiation. From detailed studies of the basic properties of this semiconducting layer it is shown that even better thermal insulation is possible. The necessary characteristics of the layer can be realized by a proper control of both the coating process and the subsequent reduction process. This has been realized not only on a laboratory scale but also industrially, thus opening the way to a new range of low pressure sodium lamps with a higher luminous efficacy and a lower power consumption.

The characteristics of the new range of energy saving SOX-E lamps will be presented, using the SOX-E 131 lamp as an example. This lamp has an efficacy of 200 lm/W. Its performance on existing, as well as on new more efficient control gear will be discussed.

n° 12

THE INFLUENCE OF THE WALL TEMPERATURE ON THE LUMINOUS EFFICACY OF HIGH-PRESSURE SODIUM LAMPS

J.A.J.M. van Vliet and J.J. de Groot
Nederlandse Philips Bedrijven B.V., Lighting Division
Eindhoven
The Netherlands

For an optimum design of high-pressure sodium lamps it is important to know the relationship between wall temperature and luminous efficacy [1,2,3].

At higher wall temperatures the modified energy balance leads to an increase of the radiation efficiency due to lower conduction losses. Moreover the spectral power distribution will be better adapted to the eye sensitivity curve because, for a constant electric power, the axis temperature will be lower at higher wall temperatures. The increase in the radiation efficiency at higher wall temperatures can be estimated by solving the energy balance equation. The luminous efficacy of the radiation can be derived from the calculated spectral power distribution. From these calculations it is concluded that, for a constant electric power, the radiation efficiency as well as the luminous efficacy of the emitted radiation both will increase by about 2% per 100K increase in wall temperature, for temperatures between 1500 and 2000K. The total effect will therefore be 4% per 100K.

Experimental verification of the calculated influence of the wall temperature can be realised by a reduction of the tube wall thickness. This leads to a lower wall emissivity. Thus a thinner tube wall gives rise to a higher wall temperature.

However the reduction of the wall thickness has an extra positive effect on the luminous efficacy as the transmission of polycrystalline alumina increases and the reflection decreases. In fig. 1 the measured relative increase in luminous efficacy of a series of low wattage high-pressure sodium lamps is given as a function of the measured wall temperature. Taking into account the improved optical properties the measured increase in the luminous efficacy with wall temperature diminishes from 7 to about 5.5% per 100K which is slightly higher than the calculated effect.

1. Waymouth, J.F., and Wyner, E.F.,
J. Illum. Eng. Soc., vol. 10, pp. 237 - 244, 1980.
2. van Vliet J.A.J.M., and de Groot, J.J.,
Frühjahrstagung der Deutschen Physikalischen Gesellschaft,
Würzburg, Germany, 1982.
3. Zollweg, R.J.,
IES Conference, U.S.A., 1982.

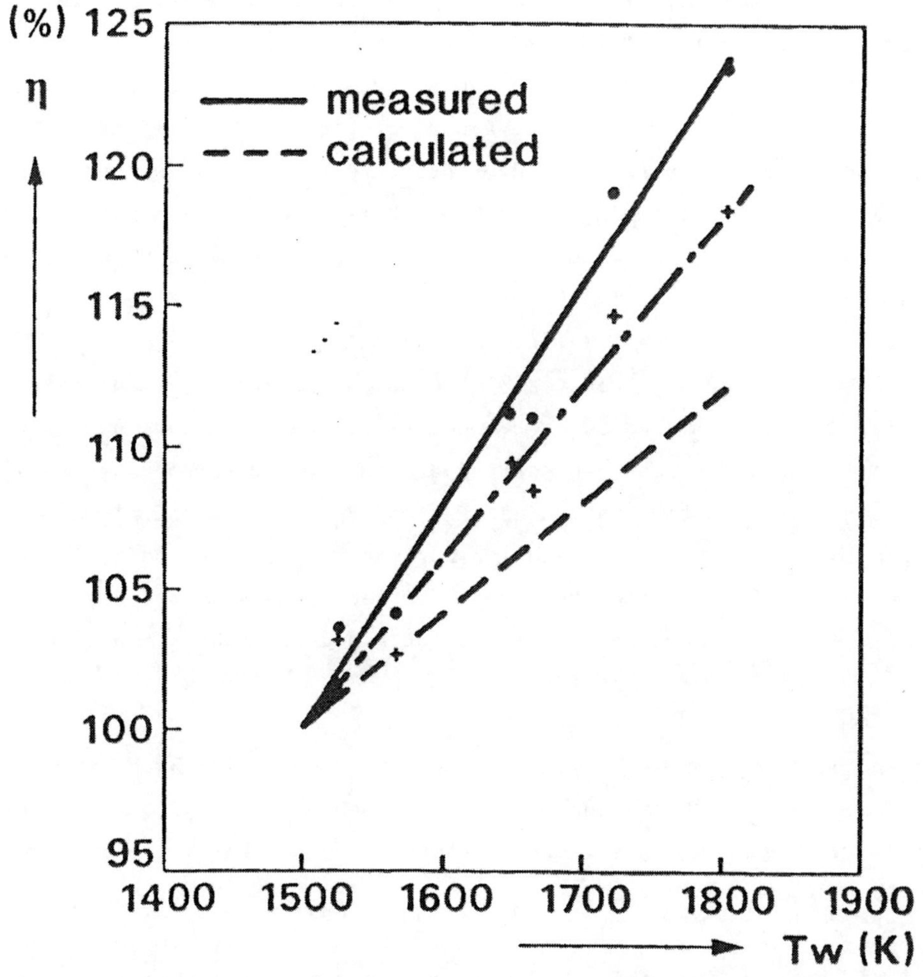

Fig. 1.

The calculated and measured relative increase in the luminous efficacy, η, of a pure sodium 70W HPS lamp as a function of the wall temperature, Tw, realised by varying the wall thickness between 0.3 and 0.6 mm.

——— measured values

-.-.-. measured values with correction for the changed optical properties

------ calculated values

n° 13

DETERMINATION OF THE ELECTRIC FIELD STRENGTH IN THE HIGH PRESSURE SODIUM LAMP

ZHOU Tai-Ming*, SHEN Ji-Ping**, WANG Jian-Rong** and CAI Zu-Quan*
* Research Institute for Electric Light Sources, Fudan University, Shanghai, China.
** Yaming Lamp Works, Shanghai, China.

In designing the HPS lamp, it is necessary to know its axial electric field strength E_z. Some arc models[1] of the HPS lamp have been founded successfully. However, in order to fit the calculations to measurements, artificially some correct factors had to be used. In this study, we measured and also calculated E_z in the HPS lamp. By adopting the data of the operating sodium pressure P_{Na} and mercury pressure P_{Hg} calculated from Hirayama's[2] formula, and taking the lowering of ionization energy of the HPS plasma into account, we found that the calculated value of E_z was very near to the measured value.

The inner diameter (ID) of arc tube was 8 mm. The gap lengths between electrodes were 50 mm, 60 mm, 70 mm, 80 mm and 90 mm respectively. The lamps were burned vertically. A heating coil around the lower end of the arc tube was used to control the coldest spot temperature T_{cold} of the tube, hence P_{Na} and P_{Hg}. The up end of the arc tube was wrapped with molybdenum foil. The heating current was so adjusted that the spectral power distribution (SPD) of the HPS lamps were kept constant in cases of different arc length at the same current of 3 A. The E_z obtained from above experiment was 15.6 V/cm.

Then the lower end of the arc tube was submerged into an indium bath. At the same SPD T_{cold} was measured by a thermocouple. In this case, T_{cold} was 990 K, the mole fraction of sodium in the amalgam was .72. Therefore we found that P_{Na} = 85 Torr and P_{Hg} = 516 Torr from Hirayama's formula.

As mentioned by Elenbaas and Waymouth, the two-temperature step profile is useful for simplifying the analysis. According to this model, $T = T_{eff}$ for $0 \leq r \leq fR_w$ and $T = T_w$ for $r > fR_w$, where R_w is the inner radius of the arc tube, f is a factor less than unit, T_{eff} is the effective temperature of arc column, T_w is the wall temperature of arc tube. The formula to calculate f was given by Waymouth. In this case we obtained T_{eff} = 3400 K by measuring the intensity ratio of line 5154 Å to line 4983 Å.

The E_z is given by the integrated Ohm's law.

$$E_z = I.\left[P_{Na}/\mu_e(Na) + P_{Hg}/\mu_e(Hg)\right]/2\pi(fR_w)^2 e\ n_e \qquad (1)$$

where μ_e(Na) and μ_e(Hg) are the electron mobility of sodium and mercury at 1 atm. We adopted the value of μ_e(Na) reported by T. S. Jen and the value of μ_e(Hg) obtained by Morrie. The electron density can be determined by Saha's equation.

Putting the value of P_{Na} and T_{eff} into Saha's equation, we obtained n_e = 1.69 $\times 10^{15}/cm^3$. However, in such a dense plasma, it is necessary to consider the effect of lowering of ionization energy of the plasma. We used Drawin's[3] data and found that ΔE_i = .143 eV. At this moment, Saha's equation must be corrected, and that is

$$n_e = 3.55 \times 10^{19} \left(\frac{T_{eff}}{5000}\right)^{1/4} P_{Na}^{1/2} e^{-(E_i-\Delta E_i)/2kT_{eff}} \qquad (2)$$

From this equation, the electron density was resolved to be 2. 18 $\times 10^{15}/cm^3$. Therefore the calculated value of E_z was 16.3 V/cm.

The partial pressure of sodium and mercury in HPS lamps were estimated with spectroscopic methods before. It was said that the shift of blue wing maximum of D-line $\Delta\lambda_B$ was proportional to P_{Na}, while P_{Hg} was relative to both $\Delta\lambda_B$ and $\Delta\lambda_R$ the shift of red wing maximum of D-line. Chien's[4] formula was one of those spectroscopic formula. Hirayama's formula was based on chemical methods. An experiment has been done to see which is better, Hirayama's formula or Chien's. Taking the value of $\Delta\lambda_B$ and $\Delta\lambda_R$ from a tube with ID of 7.5 mm and the mole fraction of sodium of .57 at T_{cold} = 850 K and I = 4A. We found that the results from Hirayama's formula were P_{Na} = 5.5 Torr and P_{Hg} = 254 Torr, and those from Chien's were P_{Na} = 11 Torr and P_{Hg} = 2780 Torr respectively. It is obvious that their results are quite different. If we used Chien's formula, the calculated value of E_z would be much larger than the experimental one. There is no doubt that Hirayama's formula is superior.

REFERENCES

1. J. F. Waymouth, J. Illum. Eng. Soc., Vol. 6, n° 3, 1977, p. 131.
2. C. Hirayama et al., "The Vapor Pressure of Sodium and Mercury over Sodium Amalgam at HPS Lamp Operating Temperatures."
3. H. W. DRAWIN and P. FELENBOK, Data for Plasmas in Local Thermodynamic Equilibrium, Paris, 1965.
4. Chien Yu-Min, Journal of Applied Physics, Vol. 51, n° 6, 1980, p. 2965.

n° 14

ENERGY BALANCE FOR HIGH-PRESSURE SODIUM LAMPS

by D O Wharmby
Research and Engineering Laboratories, THORN EMI Lighting Limited, Melton Road, Leicester, England.

The high-pressure sodium lamp has become a mature product of outstanding commercial importance. It is important now to look at its limitations and to see if there are ways of overcoming them. Fundamental information is contained in the 'energy balance' of the lamp: this is as detailed a breakdown as possible of the processes within the lamp which dissipate power.

Measurements have been made of the absolute spectral power distribution (300-2500nm) of high-pressure sodium discharges for powers 300 to 600W. The lamps used had synthetic sapphire arc tubes, internal diameter 7.4mm with arc length 80mm. Sodium mole fractions around 0.7 were employed and the tubes were filled with 5 torr (cold) xenon. The thermal emission spectra from the arc tube were measured by observing the decay of signal after switch off at wavelengths where there were no strong lines. The magnitude of the continuum radiation from the discharge (850-2500nm) was estimated by subtracting thermal and line contributions from the whole. Measurements were also made of maximum arc temperature and wall temperature.

The figure shows typical results for the energy balance. The residual power is that which was not measured. There is good agreement with the results of Jack and Koedam (1974) except that I find rather more IR radiation.

The spectral intensities of the D lines agree well with those obtained from energy balance calculations by Jones and Mottram (see Wharmby, 1980) even though the calculations over estimate maximum arc temperature by ≈300K. The calculated energy balance disagrees with the measurements, principally because of omission of some infra-red lines and approximate treatment of D line broadening.

The IR continuum is of considerable interest. Waymouth (1975) finds 17% of the input power is dissipated in the continuum. About half of this is at wavelengths shorter than 850nm and includes far wing D line radiation and recombination continua whose origin is understood. At wavelengths longer than 850nm he finds ≈9% of the input power is radiated as continuum: my measurements show 8.6 ± 0.5%. However, half of this (4.1 ± 0.5%) is thermal radiation from the arc tube, leaving 4.5% as true discharge continuum. This is the only part of the spectrum which is not understood and so cannot be included in the energy balance calculation.

Jack, A.G. and Koedam, M., 1974, J.I.E.S., 3, 323-329
Waymouth, J.F., 1975, 'An elementary model of the high-pressure sodium lamp', paper read at First Incoherent Light Sources Symposium, Loughborough
Wharmby, D.O., 1980, IEE Proc., 127A, 165-172

Fractions of input power dissipated by various processes

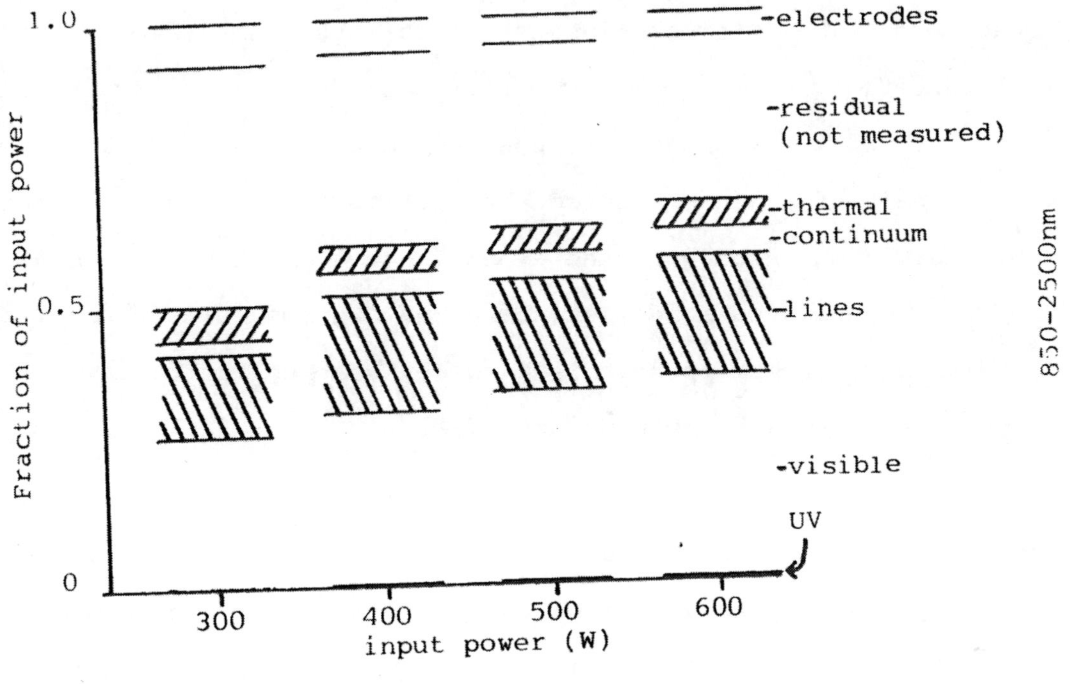

n° 15

MEASURED AND CALCULATED VARIATIONS OF EFFICACY WITH INPUT POWER PER UNIT LENGTH IN HIGH PRESSURE SODIUM LAMPS

P L Denbigh, B F Jones, D A J Mottram

Research & Engineering Laboratories, THORN EMI Lighting Limited, Melton Road, Leicester, LE4 7PD, England.

High pressure sodium lamps are designed by determining values for eleven variables (Denbigh,1977). In this process some variables are specified, some fixed and some constrained. If any of these restrictions can be relaxed then a whole new generation of lamps becomes feasible. For example the advent of electronic ballasts may extend the range of lamp voltages which are possible for any given power input. This would affect many of these variables, especially the electrode separation. The effect on arc efficacy of changing the electrical power dissipation per unit length of arc has been predicted using a computer model (Eardley et al,1979; Jones and Mottram,1981). These predictions have been checked by experiment.

We have investigated a discharge contained in a cylindrical, polycrystalline-alumina arc tube with a tungsten electrode sealed in at each end. The internal diameter of the tube was 7.4mm, and the distance between the electrode tips was 80mm. A reservoir containing 800mg of sodium amalgam was appended to one end of the arc tube. The large mass of amalgam ensured that its mole fraction was maintained at 0.74Na:0.26Hg throughout the experiment, even though the sodium and mercury evaporated at different rates. We controlled the pressures of sodium and mercury in the discharge by heating the reservoir in an indium bath to a temperature of 973K.

We used the data of Hirayama et al (1981) to estimate the pressures of sodium (about 160torr) and mercury (about 325torr) at 973K; the operating pressure of xenon was 175torr. These pressures formed the starting conditions for the calculation of the plasma composition as a function of temperature. We calculated the energy balance for an infinitely long positive column of the high pressure sodium discharge at specified dc electrical power inputs of $3589Wm^{-1}$, $4775Wm^{-1}$ and $5959Wm^{-1}$. These corresponded to powers of 300W, 400W and 500W in the experimental lamp given that the electrode voltage drop was 4V (Denbigh and Wharmby,1976). The wall temperature was assumed to be 1500K. The calculated efficacy for a power dissipation of $4775Wm^{-1}$was $166lmW^{-1}$; the other efficacies were normalised to this one. The calculated relative efficacy of the arc increased by $0.16\pm.03\%$ per arc watt per cm at constant wall temperature.

The relative arc efficacy was measured for power inputs of 300W,400W and 500W. Variations in power affected the wall temperature which in turn affected the efficacy. To overcome this problem, measurements were made at several different wall temperatures for each power input; the efficacy at 1500K was obtained by interpolation. The wall temperature was controlled by passing inert gases with different thermal conductivities over the arc tube. The measured relative efficacy at 400W and 1500K wall temperature was taken to be 100% as it was in the calculations. The relative efficacy increased at a rate of $0.35\pm.03\%$ per arc watt per cm at a constant wall temperature of 1500K.

The theoretical and experimental results thus agree to about a factor of two; this agreement is considered to be good. In contrast,Waymouth and Wyner (1981) predicted that the relative efficacy <u>decreased</u> at a rate of 0.5% per arc watt per cm.

Denbigh, P L , 1977, LR&T, 10, 28-32.
Eardley G,Jones B F,Mottram D A J, Wharmby D O,1979,J.Phys D,12,1101-15
Waymouth J F,Wyner E F,1981,J. Illum. Eng. Soc., 10, 237-44.
Hirayama C,Andrew K F,Kleinosky R L,1981,Thermochimica Acta,45,23-27.
Denbigh P L,Wharmby D O,1976,LR&T, 8, 141-5.
Jones B F,Mottram D A J,1981, J. Phys. 14, 1183-94.

n° 16

OPTO-GALVANIC DETECTION OF ACOUSTIC RESONANCES IN A HIGH PRESSURE SODIUM DISCHARGE

M.J. Jongerius and A.J.M.J. Ras, Philips Research Laboratories, Eindhoven, The Netherlands

As reported by several authors[1,2], arc instabilities in high pressure discharges may occur due to acoustic resonances inside the discharge vessel. Such an acoustic resonance can be excited when a modulated power is supplied with a frequency in the vicinity of the resonance frequency. The visible arc instability shows a threshold--like behaviour as a function of the modulation depth.

In a high pressure sodium discharge ("SON"), we have observed experimentally the presence of longitudinal acoustic resonances well below the discharge instability threshold. We have used a number of methods, the most important of which is the opto-galvanic effect detection[3]. A dye-laser was tuned to about 5835 Å, which is just outside the self-absorption minimum in the emission of the Na-D lines. The laserbeam was chopped. The discharge was burning in a cilindrical sapphire tube on a DC (commuted) supply current and was irradiated by the laserbeam at three axial positions, respectively (see Fig. 1). The amplitude and phase shift of the laser-induced changes in the voltage across the discharge electrodes were measured as functions of the chopper frequency. Experimental results are presented in Fig. 1. Effects of the first and second longitudinal acoustic resonance are clearly detected at about 2 and 4 kHz, respectively. The excitation of the acoustic resonances strongly depends on the axial position of the laserbeam, because by irradiation at a knot in the standing acoustic wave no excitation can occur. Indeed, we did not find the first longitudinal mode at laser position II (see Fig. 1). The non--vanishing excitation of the second longitudinal mode measured at positions I and III may be explained by the low gastemperature behind the electrodes, which displaces the knots in the acoustic wave.

An approximate discharge model will be presented which qualitatively describes the opto-galvanic measurements.

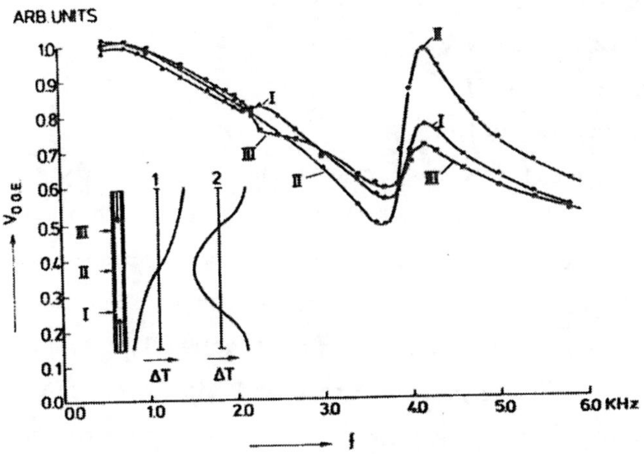

Fig. 1. The measured amplitude and phase shift as functions of the chopper frequency. I, II and III mark the axial positions of the laserbeam.

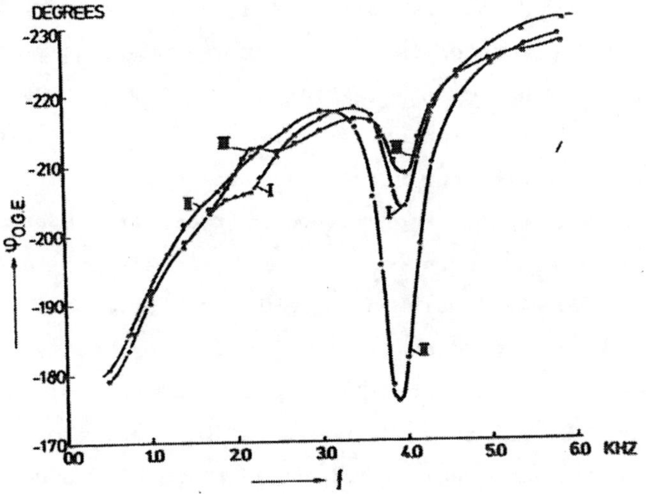

References

1. C.F. Gallo and W.L. Lama, Appl. Opt. 16, 819 (1977).
2. H.L. Witting, J. Appl. Phys. 49, 2680 (1978).
3. W.J. van den Hoek and J.A. Visser, J. Appl. Phys. 51, 5292 (1980).

n° 17

THE KINETICS AND CONSEQUENCES OF ALKALI PERMEATION THROUGH HPS ENVELOPES

by

W. H. Rhodes, D. L. Dugger, and A. E. Feuersanger
GTE Laboratories
40 Sylvan Road
Waltham, MA 02254

High pressure sodium lamps commonly employ a polycrystalline Al_2O_3 envelope which is generally considered to be an effective container for Na vapor. Occasionally, glass outer jacket coloration is observed either in short of long life HPS lamps; this condition is accentuated when high wall temperatures are employed. Several investigators attribute this coloration to Al deposition from incongruently vaporizing Al_2O_3. One study demonstrated that Na can permeate both poly and single crystal Al_2O_3 envelopes under high electric field gradients which suggested an alternative explanation for the outer jacket coloration.

This study was undertaken to delineate Na permeation through polycrystalline Al_2O_3 under conditions which simulate the environment in an operational HPS lamp. Limited experiments covered polycrystalline Y_2O_3 and single crystal Al_2O_3 HPS lamps. The unjacketed lamps were mounted in a 10^{-9} torr vacuum system adjacent to a quadrupole mass spectrometer. The lamps were biased to -70 Vdc to balance the potentials associated with the quadrupole ion source. It was found early in the investigation that K as well as Na permeated the arc tubes. The peak height at Na_{23}^+ and K_{39}^+ were monitored as a function of lamp wattage by repetitive scanning using a MINC 11/23 computer. A standard production lamp designed to operate at 150 W was operated over the range of 125 to 265 W using a variable reactor. Wall temperature was measured independently by two techniques; IR optical pyrometry and the radiance Na D-line center method. The arc tube wall temperature range of 1080°C to 1300°C was covered. The internal Na pressure was monitored by measuring the Na D-line separation. Several reservoir lamps were constructed to permit independent control over wall temperature and Na pressure.

The NaHg amalgam contained 0.1% K, which diffused markedly faster than Na in polycrystalline Al_2O_3. Both species diffuse by processes that are described by exponential-activated process equations. Some of the data showed evidence for two activation energies typical of diffusion data. Two mechanisms operate which are dominant in different temperature regions. The calculated diffusion coefficients are compared with Na tracer diffusion measurements in Al_2O_3. The permeation rate depends on the alkali pressure, as demonstrated by comparing rates for the standard and reservoir lamps, and by adjustment of the pressure through independent heating of the reservoir.

The darkened Pb glass outer jackets from several lamps operated at >1250°C wall temperature were found to have an Na rich inner surface layer by both ESCA and Auger analysis. It is theorized that Na reduces the Pb^{+2} in the glass, allowing charge transfer processes between oxidation states which are responsible for the light brown to orange color of the lead glass. Several special lamps containing fused quartz or Pyrex sleeves around the ceramic envelope also exhibited Na deposition after extended lamp operation at high temperature. These glasses were not colored, presumably due to the absence of PbO. This study clearly demonstrates the kinetics of alkali permeation through operating ceramic envelopes, which we believe explains outer jacket darkening in HPS lamps.

n° 18

A STUDY ON THE TEMPERATURE DISTRIBUTION IN HIGH-PRESSURE SODIUM ARC

WU Mei-Ping, ZHOU Tai-Ming, CAI Zu-Quan
Research Institute for Electric Light Sources, Fudan University, SHANGHAI, China

The method of treating radiative transfer is of much importance for calculating temperature distribution, particularly in the optically thick plasma, such as high-pressure sodium discharge which will be discussed in this paper, not only the emission of radiation, but the absorption of radiation exists. The treatment of radiative transfer is much trouble in such a plasma. Consequently the purpose of our work is to advance another method for simplifying this problem, which treats the radiation transfer with diffusion approximation as follows :

$$\overline{F_{R\nu}} = - \frac{4\pi}{3K_\nu} \nabla J_\nu$$

where $\overline{F_{R\nu}}$ the radiation flux per unit frequency, K_ν the absorption coefficient at frequency ν, J_ν the average radiation intensity.

We used the diffusion approximation to calculate the temperature profile for the sodium arc having I = 4.5 amps at 0.33 atom and found that the result agreed well with that obtained by Waszink and De Groot.[(1-2)] In the calculation we took account of energy losses due to the thermal conduction and radiation, and applied the trial place method and the trial value. To correct the effect of the radiation from several strong lines other than Na-D line, i.e. 568.3/568.8 nm, 818.3/819.5 nm, 1138.1/1140.4 nm, as modified index was introduced in this paper.

We also measured the temperature distribution with experiments for verifying the calculated result. Because of the high-pressure sodium plasma is optically thick and self-absorption of the spectral lines can't be neglected, the temperature distribution may be measured by using the spectroscopic procedure developed by Göing and Bartels.[(3-4)] The obtained axial temperature is in agreement with our theory calculation.

It is concluded that the diffusion approximation is an accurate and simple method for studying the high pressure and optically thick sodium discharge.

REFERENCES

1. J. J. De Groot and J. A. J. M. Van Vliet, J. Phys. D : Appl. Phys., vol. 8, 1975.
2. J. H. Waszink, J. Phys. D : Appl. Phys., vol. 6, 1973.
3. N. Ozaki, J.Q.S.R.T., vol. 11, 1971.
4. H. Zwicker, "Plasma Diagnostics", chap. 4, 1968.

n° 19

MEAN AND LOCAL MODULATIONS OF THE SPECTRUM EMITTED BY HIGH PRESSURE MERCURY THALLIUM DISCHARGES WITH DIFFERENT MERCURY AND THALLIUM PRESSURES

Jean-Jacques DAMELINCOURT, Stylianos KOURIS, Michel AUBES and Jacques SALON
Centre de Physique Atomique, Laboratoire Associé au CNRS n° 277, Université Paul Sabatier, 118 route de Narbonne, 31062 TOULOUSE CEDEX, France
Godefroy LEMINTIER
Centre Technique de la Compagnie des Lampes, 11 Quai de Dion Bouton, 92818 PUTEAUX, France

In discharge lamps operated at 50 Hz, temperatures and electrons density distributions vary with time during a cycle of the electric field (1,2). Spectral emission is modulated because of these variations. Several authors (3,4) have studied line emission evolution with time but few (5) have investigated additive influence on discharge emission. In this work, we intend to study the effect of mercury and thallium partial pressures on emission and modulation in visible and U.V. spectral regions for a high pressure mercury discharge containing thallium iodide.

We can adjust additive pressures by moving a crucible which surrounds the lower part of the lamp. In this way, for lamps where thallium iodide is not completely vaporized, we can, for a given a.c. power or a given a.c. current, change cold spot temperature and so thallium pressure.

The thallium pressure is measured by comparing the 535 nm thallium line profile with the one emitted by a reference lamp already calibrated. Mercury pressure and temperature profile are determined simultaneously from side on measurements of the absolute intensity of the mercury 577 nm line using the Abel inversion technique.

The modulation of the more intense lines of the discharge was measured for the central part of the dicharge and for the space averaged instantaneous spectral intensity.

The temperature modulation decreases when mercury pressure increases. Corresponding to this lower modulation one finds a lower modulation for the line in the discharge with higher mercury pressure.

In fact the temperature modulation interacts with the radiation trapping : a discussion of the effect of the both phenomena is given.

Visible over ultra violet energy ratio is evaluated at two instants of the period for the central part of the discharge and for the space averaged instantaneous intensity. The results of figure 1 show that the visible/UV emission is a function of both the additive and mercury pressure. This ratio strongly varies with time.

The central part of the lamp gives a modulation lower than the one obtained from the mean space values. Probably this is mainly due to the exponential dependance of the densities populations versus temperature which gives for a given level a stronger variation for lower temperatures.

Knowing the modulations of the visible and ultra-violet part of the spectrum for various pressures of mercury and thallium, it is possible to evaluate the modulation of the output light in lamp corrected with phosphors.

REFERENCES

1. J. J. LOWKE, R. J. ZOOLWEG, R. W. LIEBERMANN, J. Appl. Phys. 46, 650 (1975).
2. J. J. DAMELINCOURT, D. KARABOURNIOTIS, L. SCOARNEC, P. HERBET, J. Phys. D : Appl. Phys., vol. 11, p. 1029 (1978).
3. J. J. DAMELINCOURT, D. KARABOURNIOTIS, M. KARABOURNIOTIS, L. SCOARNEC, J. Phys. D : Appl. Phys., vol. 11, p. 2207 (1978).
4. H. P. STORMBERG, J. Appl. Phys. 31, 1963 (1980).
5. C. F. GALLO, Appl. Optics 5, n° 8, 1285 (1966).

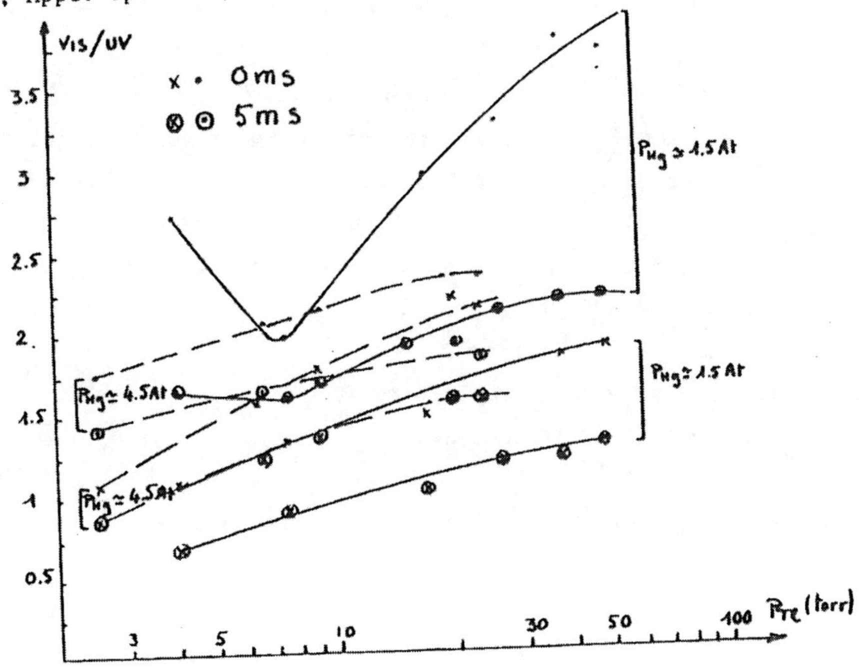

FIGURE 1 : Visible (Tl 535 nm + Hg 546 nm) over ultra-violet (Tl 356 nm + Hg 365 nm + Tl 377 nm) ratio.

n° 20

EXPERIMENTAL SET UP FOR THE STUDY OF THE ELECTRICAL CHARACTERISTICS OF HIGH PRESSURES DISCHARGES SUPPLIED BY CURRENT OR VOLTAGE SOURCES DRIVED BY MICROPROCESSOR

Jean-Jacques DAMELINCOURT, Bernard BELLON, Jacques SALON and Dominique GUILHEM
Centre de Physique Atomique, Laboratoire Associé au C.N.R.S. n° 277, Université Paul Sabatier, 118 route de Narbonne, 31062 TOULOUSE CEDEX, France

This work is supported by the Association Française de l'Eclairage.

The study of the electrical properties of the discharge and of their dependence on physical parameters become more and more important with the expansion of electronic lamp devices.

The actual project is concerned with an experimental set-up designed for three kind of studies :

- The effect of discharge composition on the static and dynamic characteristics of the discharge.
- The impulse response of the discharge to voltage or current square wave excitations. Either for low or large signals. This point is of interest because it can lead to the time constants characteristic of the discharge.
- The effect of current deformations on the line emission of the discharge.

The experimental set-up will be composed of :

1) A source system (figure 1) fit in order to work as a current source or as a voltage source. In the two cases the sources system may be drived by a microprocessor. This microprocessor allows the choice of the voltage or current characteristics : frequency, on and off duration, range of current and voltage, shape of the wave form.

2) A measuring system including :

- a high resolution monochromator,
- a photomultiplier with a sample circuit,
- an optical multichannels analyser.

This experimental set-up is illustrated in figure 2.
(1) step motor ; (3) arc carrier ; (4) lamp ; (5) retractable mirror ; (6) lens ; (7) monochromator ; (8) frequence source ; (9) photomultiplier and (10) its associated electronic ; (11) microprocessor ; (12) gratting command ; (13) printer ; (14) (15) storage devices ; (16) microprocessor command ; (17) calculator.

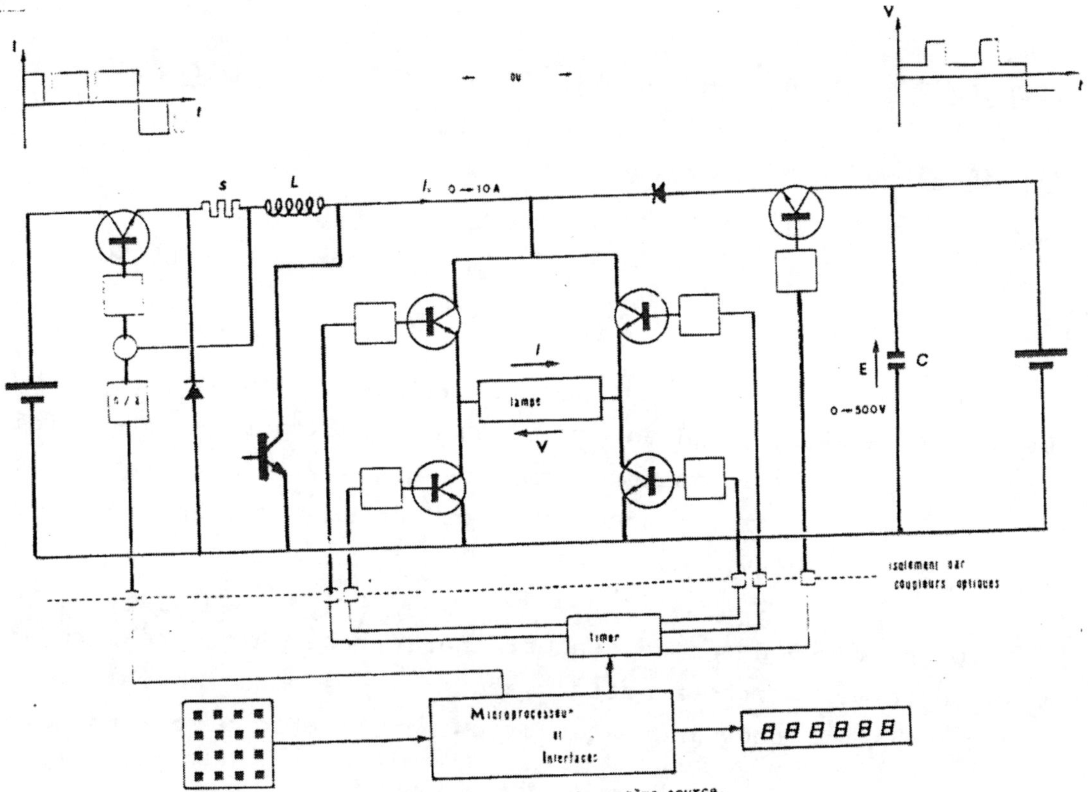

FIGURE 1 - Schéma bloc du système source.

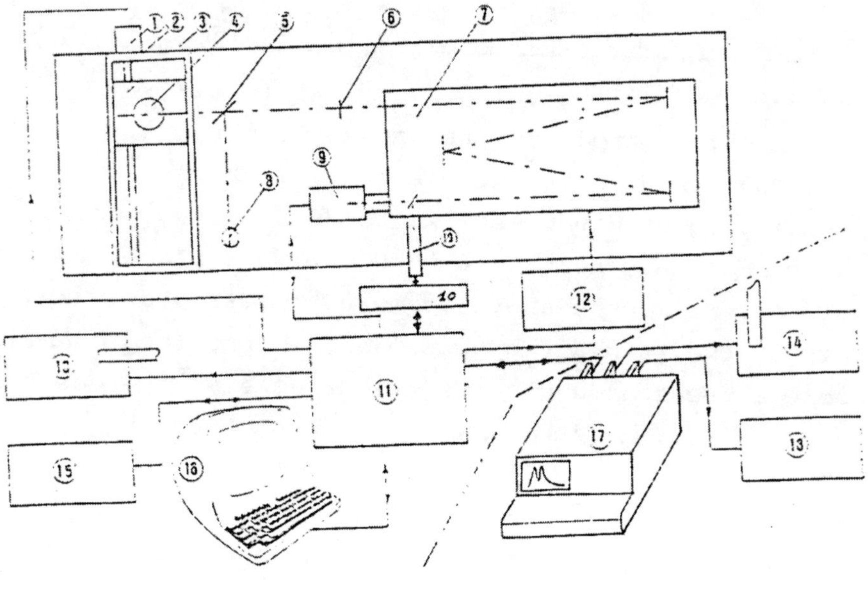

FIGURE 2 : Schéma bloc du dispositif expérimental.

DISCHARGE HIGH PRESSURE IN THE MERCURY AND METAL HALOGENS

APPLICATIONS

SURFACE TREATMENT, INK DRYING BY PHOTOPOLYMERIZATION

Author : J. VILLAIN
CDL - MAZDA

Device consists in photopolymeriziring resins, inks, pigmented, photo-sensitive products, and so on ... on every support treated ordinary by varnishings, inks, pigmented products mixed with evaporable solvents and, because of this, producing power consumptions and occupying great volume as also heavy investments.

Device includes one or several discharge tubes high pressure stabilized by transformers with magnetic reactance as also a control and transfer device. The efficacy increases by using elliptic or parabolic reflectors.

The lifetime tubes fluctuates between 1 500 and 2 000 hours approximate. The productivity increases appreciably.

INTEREST OF THE TECHNIC UV APPLYED TO VERNISHINGS, INKS, PROTECTIVE AND SO ON ...

For the classical products the drying is obtained by the evaporation of solvents contained in the vernishings, inks, and so on ... Out of 100% of initial product used, from 20 to 30 % only of useful dryed extract will be effectively kept on the receptive surface, the rest, therefore the solvent, will be evaporated more or less slowly and wasteful with the help of the warm air by sweeping, in the long tunnels by example, therefore a great ground surface lost as also the energie wasted uselessly.

This device, in opposition to classical device by evaporation, doesn't require heating but only one or several spectrum radiations UV centred on the absorption curve of photoinitiators retained (benzoin ether, benzophenol, and so on ...).

Picture I

efficiency at 1000 hours compared with 100 hours

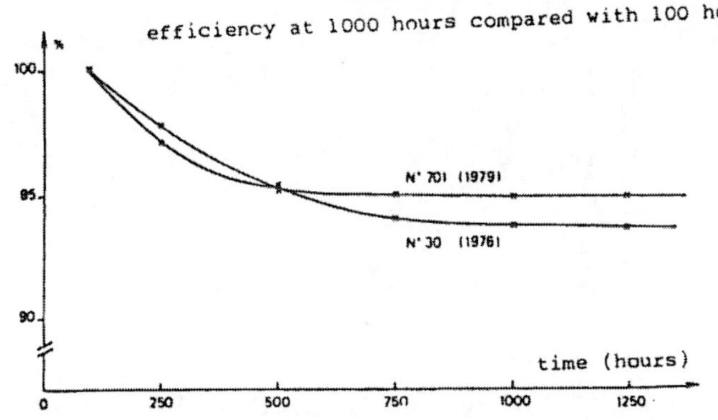

energy at 100 hours

tube 55-200 ———

other tube ------

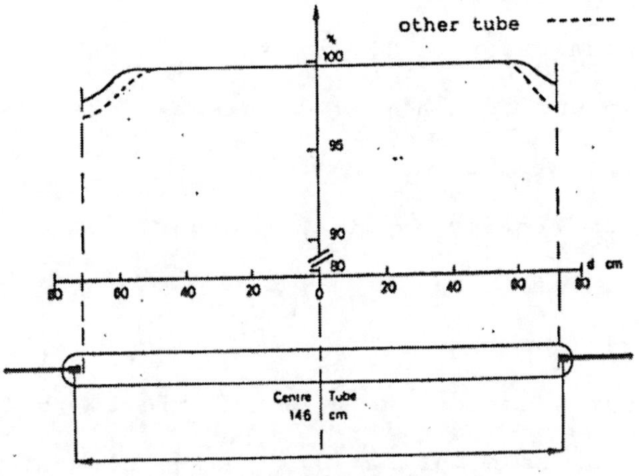

Picture II

flow distribution at 100 hours tube 146 cm

energy at 1000 hours

tube 55-200 ———

other tube - - - -

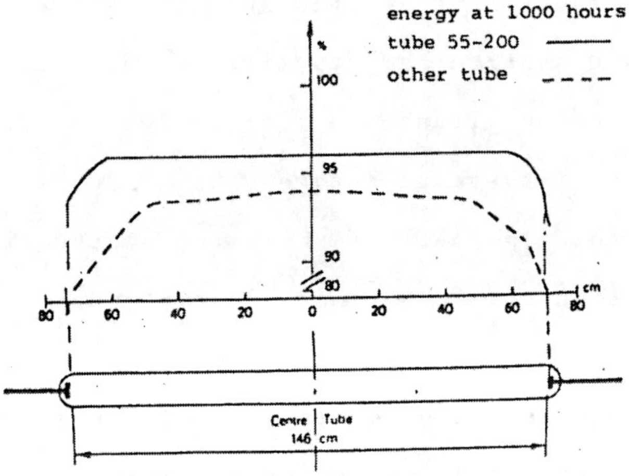

picture III

flow distribution at 1000 hours

ELECTRONIC BALLAST FOR GAS DISCHARGE LAMPS

J. Rozenboom

Nederlandse Philips Bedrijven B.V., Lighting Division

Eindhoven

The Netherlands

The feasability of several combinations of gas discharge lamps and electronic ballasts will be discussed. The electronic ballast has the following functions:

- suitable for connection to the mains.
- conversion of power to a higher frequency (preferably above 20 kHz, the audible noise limit).
- current or power limitation at the output, for direct connection of a gas discharge lamp.
- ignition of the lamp.
- sufficient suppression of the adverse effects on the environment due to the techniques used, primarily - mains current distortion
  - electromagnetic interference.

The electronic ballast, as defined above, combined with the appropriate gas discharge lamp is aimed at producing light at low energy cost or low applicance cost or low weight or small dimensions or a combination of these advantages for a special purpose.

There are two applications where the electronic ballast has been a feasable solution for long time: firstly where only a dc voltage is present as in trains, busses, boats and cars; and secondly where only very high frequencies are possible, the electrodeless discharge lamps as applied in the field of atomic spectroscopy as intense atomic line sources.

Recent developments have been triggered by the rise in the cost of energy. The main areas where electronic ballasts are capable of conserving energy are:

1. The lamps produce more light if fed with energy at high frequency. This phenomenon is observed with fluorescent lamps.
2. The ballast produces less heat than its equivalent copper-iron counterpart.
3. Dimming of the artificial light level causes energy saving e.g.
   - office lighting, in combination with daylight (energy management)
   - street lighting where the illumination is a function of traffic density.

Thus if good dimming properties are required electronic ballasts have many advantages.

The availability of electronic components also plays a major role in the development of electronic ballasts. Certainly the circuits for switched mode power supplies (SMPS) and line deflection in television sets have stimulated the transistor makers to develop suitable devices like bipolar transistors of up to 1600V and mosfet's of up to 500V. Their function in electronic ballasts is very similar to the above mentioned applications. The same development can be observed in the case of passive components like magnetic materials and capacitors.

An example will be given of the development of an electronic ballast.

n° 23

EXPERIMENTAL AND THEORETICAL STUDY OF COMPENSATED TRANSFORMERS FOR DISCHARGE LAMPS. ANALYSIS OF THEIR INFLUENCE ON THE EMISSION OF THE LAMPS.

Ara SIMONIAN
Polytechnique Institute of Erevan, 105 rue Terian-Erevan-9, EREVAN 375009, USSR

The specific case of a great number of technological processes requires the rise of high power discharge lamps with an emission spectrum of particular composition.

The operation of such lamps with alternative voltages often necessitates the use of transformers or autotransformers having a high internal inductance.

This study proposes a classification of compensated transformers, points out the more adapted field of application in each case, and gives the methods for their calculation as well as the analysis of the luminous characteristics of the lamps.

The harmonic analysis and synthesis allow the obtention of analytical expressions of instant and r.m.s. values of currents, voltages and povers for each part of the circuit. For this purpose we have used the sum of the numerical and functionnal series, without limiting the number of the harmonic components.

The influence of the reactance/inductance ratio on the operating regime of each portion of the circuit, as well as the luminous characteristics of the lamps are studied. For a particular value of this ratio (2,62) the transformer, connected to the lamp through a series capacitor, functions as a current source.

The adaptation of the firing and operating regimes is obtained by solving the system of 3 equations characterizing each regime and the condition of compensation of the reactive power.

For the analysis of the influence of the current shape on the emission of the high pressure discharge lamps under thermodynamic equilibrium, a canal model of the discharge was used. Analytical expressions depending on current shape are obtained for the instant and r.m.s. values of the intensity of the lines.

It is shown that the ratio of the excitation potential over ionization potential is the general parameter defining the influence of current shape.

For a given power dissipated in the lamp, a change in the current shape may determine either an increase or a decrease of the radiated flux, depending on the value of this parameter. It is also shown that the variation of the emission for lines having the same upper level is the same.

The experimental results for mercury lamps and for metal iodide lamps show a qualitative agreement with the theoretical conclusions.

After analysis of the technological characteristics of the transformers, a simple construction is proposed for a transformer having a high internal inductance, without use of a magnetic shunt and the method for the calculation of its technological parameters is described.

n° 24

PERFORMANCES OF A NEW V.H.F. SOURCE FOR THE V.U.V. SPECTRAL RANGE

S. Girard[x], G. Levèque[x], P. Leprince[xx], J. Marec[xx] and M. Pouey[xx]
[x]Laboratoire de Spectroscopie II, U.S.T.L. Place E. Bataillon F 34060 Montpellier
[xx]Laboratoire de Physique des Gaz et des Plasmas, Université Paris-Sud, F 91405 Orsay

The main advantages of V.U.V. light sources R.F. excited are well known ; spectral purity, emission of fluctuating or not, line spectra as well as rare gases continua. To excite and sustain plasma by means of a R.F. field, several methods are available. The plasma can be created, as it is usually the case in a microwave cavity. As the resonance frequency depends on the plasma one, i.e., the electronic density and the cavity dimensions, the system is few stable and the plasma dimensions are about those of the cavity. Otherwise, the eigenmodes of the bounded plasmas can be used : either resonances (e.g. dipolar resonances), either a wave propagating along a plasma column. In the resonance case, there is a rigorous relation between the density and the frequency whereas in the propagation case, the condition is less restrictive. Our spectral lamps have been realized with this method.
A surface wave can propagate along a plasma column if :

$$f_o < f_p \, (1 + \varepsilon_g)^{-1/2} \qquad (1)$$

where f_o is the excitation frequency, f_p the plasma one (with $f_{p(Hz)} \simeq 9.10^3 n^{1/2}$, n being the electronic density in cm^{-3}) and ε_g the permittivity of the surrounding dielectric. If the coupler output power is sufficient, the relation (1) is satisfied. Then, the generated plasma is not localized in the coupler. Its length can reach a few centimeters to a few meters depending on the incident power, the dielectric internal diameter, the pressure and the gas composition.
The surfatron has been extensively described (1). Its structure has the coaxial geometry, thus the EM field in the plasma has the azimutal symmetry corresponding to the surface mode m = o.
Summarising, the main properties and performances of the surfatron as plasma generator are the followings :

- It is a wide band coupler (typically 300 MHz) with a coupling better than 90 %. Structures have been constructed for 200 to 2450 MHz.
- The plasma excitation can be obtained in continuous mode or in modulated one until about 100 kHz.
- In rare gases, pressures from 10^{-1} torr to the atmospheric pressure are allowed with the same surfatron on capillary tubes of inside diameter up to 5 mm. The gene-

ration of plasmas in molecular gases requires a higher energy and a cooling (with air or oils) is often necessary.

- At high pressure (few hundred torrs) rare gas continua have been observed. The continuous emission between 180 to 320 nm for an atmospheric argon plasma excited with a 2450 MHz microwave surfatron (V.H.F. power 130 watts, flow rate 4 ℓ/min, 5 mm tube diameter) has been compared to that of a mini argon arc working at 8 ampères ; both sources have the same radiance of the order of 0,25 mW $cm^{-2}nm^{-1}sr^{-1}$ (2).
- At low pressure (few torrs) resonance line of rares gases are strongly emitted. At the exit slit (0,1 x 10 mm) of a one meter normal incidence monochromator (horizontal and vertical aperture of 0.033), 7 10^8 ph/s have been measured at 58,4 nm (HeI, 0,75 torr), 1,4 10^9 ph/s at 73,7 nm (NeI, 0,5 torr), 9,6 10^8 ph/s at 104,8 nm (ArI, 0,1 torr) and 9,6 10^9 ph/s at 121,6 nm (HI, 90 % Ar, 10 % H_2, 0,25 torr).
- Ionic lines are strongly increased by a decrease of the pressure as well as of the tube diameter.

Actual works are related to the design of structures limiting the length of the plasma in order to increase the V.H.F. power density inside the tube.

REFERENCES

(1) J. Marec, E. Bloyet, M. Chacker, P. Leprince and P. Nghiem, Microwave discharges, NATO ASI, les Arcs, France (Juillet 1981), Plenum Press (to be published).

(2) E. Bloyet, P. Leprince, J. Marec and M. Pouey, III-65 Proceedings of the VI Intern. Conf. on Vacuum Ultraviolet Rad. Physics (Juin 1980), University of Virginia, Charlottesville (USA).

n° 25

ELECTRIC DISCHARGE LAMPS WITH MULTIPLE ELECTRODES

J.E.Harry, Department of Electronic and Electrical Engineering, University of Technology, Loughborough, Leics., LE11 3TU, U.K.

Electric discharges can be operated in close proximity to each other and supplied from the same power supply provided each discharge is separately stabilised each side of the electrodes.[1] Two modes of operation are possible; parallel, in which the current in adjacent discharges is in the same direction, causing the discharges to coalesce; and anti-parallel, in which the current in adjacent discharges is in opposite directions so that repulsive forces between them maintain them separately. Both methods of operation are applicable to high pressure and low pressure discharge lamps.

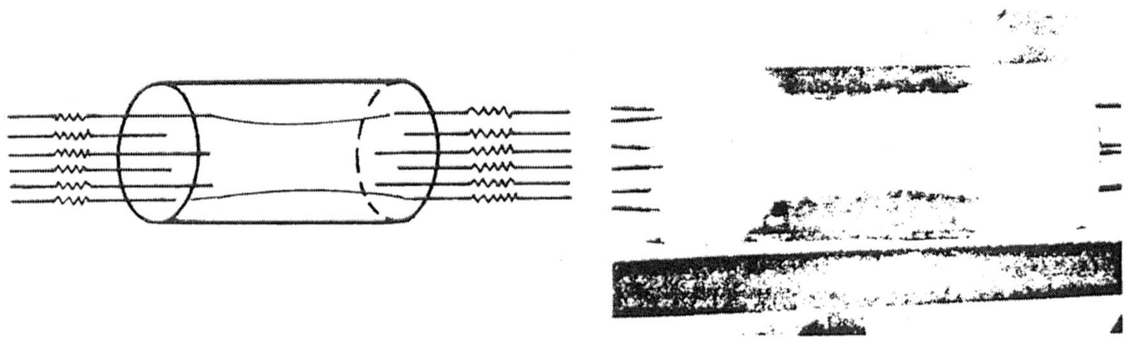

Fig. 1. Cylindrical lamp with multiple electrodes.

Fig. 1 shows a tubular low pressure cold cathode discharge lamp incorporating multiple electrodes and its stabilising circuit operating in the parallel coalesced mode for operation at a pressure of about 500 Pa in argon with mercury vapour. The total discharge current was 0.48A corresponding to an individual electrode current of 80 mA. The discharge can be seen to coalesce close to the electrodes and fills a large volume of the tube.

The use of coalesced discharges enables geometries other than tubular to be used. Fig.2 shows a rectangular cavity using three pairs of separately stabilised cold cathode electrodes operated in the parallel coalesced mode. The discharge coalesces close to the electrodes and uniformly fills the rectangular cavity. Heated cathodes have also been successfully used.

Electric arcs have been operated in the parallel and anti parallel modes in

Fig. 2. Rectangular lamp with multiple electrodes.

similar configurations. Large volumes of ionised gas can be produced in the parallel mode of operation and large arrays of discharges obtained in the anti parallel mode. The individual current loading at each electrode is correspondingly reduced, alternatively higher total currents are possible. Circular, rectangular and other geometrical configurations are also possible.

Multiple electrodes enable individual electrode loadings to be reduced or higher total currents to be used allowing either longer lamp life or higher light outputs to be obtained. Alternatively different geometries are possible not limited by cylindrical lamp envelopes. The use of multiple discharges is also being investigated for excitation of CO_2 lasers[2] and other discharge applications.[3]

References.

Harry, J.E., Knight, R., Simultaneous operation of electric arcs from the same supply, IEEE Trans., PS9, (4), 248-254, 1981.

Harry, J.E., Saleh, S.N., Multiple electrode systems for high power CO_2 laser excitation, Appl. Phys. Lett. 40, (5), 359-361, 1982.

Provisional patent application 81.08653, 08654, 08655, 08656.

n° 26

THE IGNITION OF GAS DISCHARGE LAMPS

J.A.J.M. van Vliet
Nederlandse Philips Bedrijven B.V., Lighting Division
Eindhoven
The Netherlands

Ignition of a discharge lamp - mostly initiated with the help of a starter - involves converting the starting gas between the electrodes from a non-conductive state into an arc stabilised on the mains with a given ballast. Prior to the existence of a stable burning arc the starting gas passes through a sequence of various intermediate phases:

a. The breakdown;
b. The transition from starter-pulse current to mains current operation;
c. The reignition, when the lamp current passes through zero, during the first few cycles after breakdown.

The ignition can be hampered during each of the intermediate phases as will be illustrated.

a: A rectangular starting pulse of pulse height PH_s across the electrodes of a non-conducting lamp generally causes the following (fig. 1):
At a fixed pulse voltage amplitude, voltage breakdown occurs only with pulse widths larger than a minimum value, PW_V, after which a glow or arc developes. Current breakdown can take place at a smaller width which depends on the mean statistical delay time, τ_s, and formative time lag, τ_f. A linear relationship is found between the difference of the starting pulse amplitude and minimum amplitude for breakdown - the so called overvoltage - and the reciprocal value for the formative lag time.

b: The mains current after the voltage breakdown is determined by the lamp voltage directly after breakdown and by the phase angle between the mains voltage zero and the occurrence of voltage breakdown. If the lamp current grows too slowly, or changes polarity with delay, ignition can be hampered.

c: Depending on the thermal inertia of the electrodes, electrode material, starting gas composition, pressure and power transported to the electrodes, glow discharges can occur after reignition, during several periods after breakdown. These phenomena obstruct the development of a stable burning arc.

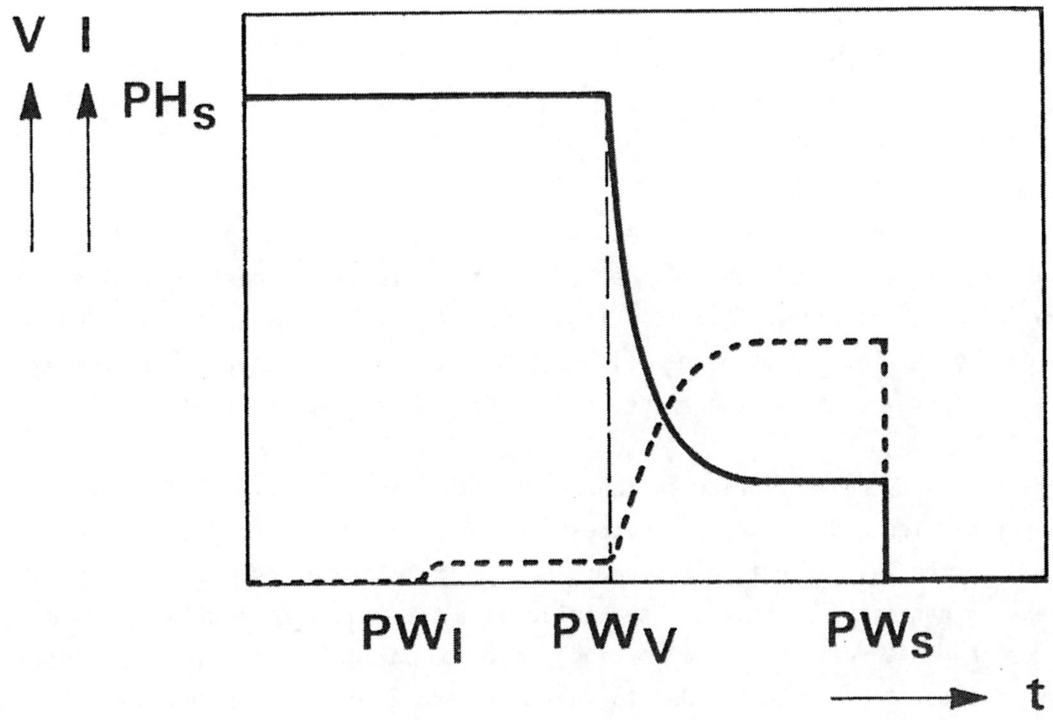

Fig. 1

Schematic showing the current, I, (dotted line) and voltage, V, (solid line) in a discharge lamp during the breakdown phase:

PW_I : the moment of current breakdown ($\tau_s + \tau_f$)

PW_V : the moment of voltage breakdown

PW_s : the end of the pulse delivered by starter

PH_s : the amplitude of the starting pulse.

n° 27

Electrochemiluminescence - a possible light source?

H. Köstlin, H. Schaper, E. Schnedler
Philips GmbH Forschungslaboratorium Aachen, Germany

Electrochemiluminescence (ECL) might offer a way of converting electrical energy into light; negative (M^-) and positive (M^+) radical ions are generated electrochemically from certain aromatic hydrocarbons (M) in organic solvents and recombine with the emission of light according to the following scheme

cathodic reaction	$M + e \longrightarrow M^-$
anodic reaction	$M - e \longrightarrow M^+$
recombination	$M^+ + M^- \longrightarrow M^* + M$
luminescence	$M^* \longrightarrow M + h\nu$.

The process is regenerative with respect to the luminescing material and the energy balance requires that the sum of the redox-potentials is at least equivalent to the energy of a photon ($h\nu$) of visible light, i.e. $\gtrsim 3$ V. The material should have a high quantum efficiency of normal fluorescence to guarantee that the system will radiate effectively after the recombination to the excited state (M^*).

The radical ions can be formed in two ways: simultaneously at two, more or less symmetrical electrodes (d.c.) or successively at the same working electrode (a.c.). In the last technique a supporting electrolyte and consequently a polar solvent is necessary which makes the solution sufficiently conductive to allow a well defined alternation of the working electrode potential against the solution by means of a counter electrode. Due to the very high reactivity of the radical ions and hence their instability, the cyclus time is limited to some 1/10 sec. The lower limit is given by the time constant of charging the electrode capacitance and is typically 1 msec. The depth of the light emitting volume is determined by diffusion of the particles during a cyclus time and was found to be about 10 μm in experiments with diphenylantracene in a solution of acetonitril with TBAP as electrolyte. The first, d.c. driven technique requires short electrode distances and at least one electrode to be transparent. ECL-cells of this type normally are found to have only a short lifetime due to side reactions with impurities. If, however, no electrolyte is added and hence no polar solvent, but an unpolar one is used, the solution can be purified much more perfectly especially with respect to water. These cells are of very low conductivity at low voltage. Nevertheless, in front of the electrodes some impurity ions can form the charged double layers whose potential drops cause the ion creating charge transfer above the threshold of the redox voltages. The weak current starting now, is due to diffusion of these ions to their counter electrodes. Above a second, slightly higher threshold the transport mechanism changes into a electrohydrodynamic convection.

This second threshold can be characterized by an electrical Rayleigh number which is related to the additional voltage driving the ions in the viscous solution. After the onset of convection the average flow velocity is given by the ratio of dielectric and kinetic energy in the fluid and increases linearly with the voltage. Consequently the electrical current and simultaneously the light output follows the voltage linearly in the convection range up to some 10 V. This was observed during experiments with thin layer cells filled with a solution of rubrene in dimethyl-oxyethane. The spacing between the transparent In_2O_3:Sn electrodes was 50 μm and the convection started at about 0.5 V above the onset of the redox process, which corresponds to a fluid velocity of about 0.2 cm/s. Due to the boundary conditions of the cell the convection has a well defined honeycomb like lateral structure (Bernard cells) which directly can be observed in the microscopic pattern of the luminescence. If the electrode itself is designed with a fine structure the regular convection is disturbed and a better mixing of the ion radicals is obtained resulting in an improved luminous efficacy.

The life time of this thin layer d.c. cell is up to some 100 h, which means that one molecule participates about 10^4 times to the ECL process. However, in both types of ECL compartements, the a.c. and the d.c. driven cell, the luminous efficacy is only about 1% and the maximum power density is of the order of 10^{-3} W per cm^2 luminescing area. The total output of light emitted by a cm^2 ECL surface during the lifetime of the cell is of the order of a few lm·h. This is very far away from the data of common light sources, but nevertheless for the adapted eye, for instance at night, ECL might be used for signals or the background illumination of liquid crystal displays, where homogeneous light density over the whole area is

n° 28

INVESTIGATION OF DISCHARGE EVOLUTION IN ARC LIGHT SOURCES

A.M.BORBAT, Z.P.KOROLYOVA, A.S.SKIRDA, V.P.TETERA

Kiev Shevchenko State University, Kiev, USSR

Alternating current (50 Hz) arc discharge in quartz-mercury-vapour lamps with tungsten electrodes made in form of rods with spiral winding and coated with emissive composition on the basis of Ba and Y oxides has been investigated.

The lamps were filled with argon up to the pressure of 530 Pa and mercury vapours.

The output of the components of the electrode emissive coating was determined through the intensity of the spectral line Ba II 455,4 nm. The character of the electrode temperature variation was investigated through the heat radiation at wave-length 650 nm. The experimental method permitted to investigate synchronously the intensity of the spectral line, heat radiation, voltage drop in the discharge, discharge current.

The investigations show that beginning from the moment of switching-on the voltage on the lamp in the time interval up to 2 s an electrical arc with cool cathode is formed. This process is accomponied by intense barium and heat radiations which are localized in the region of the cathode spots. Simultaneously the injection of easily ionizable components of the cathode coating (Ba, BaO) into the electrode area and their localization, mainly in the region of the cathode spots, take place. With that the increase of the intensity of the spectral line of Ba ions is accompanied by the decrease of heat radiation or by the slowing down of its growth.

The kinetics of the change the intensity of the spectral line of Ba ions and heat radiation during the first cathode semiperiod is illustrated by fig.I (curves I and 2, respectively).

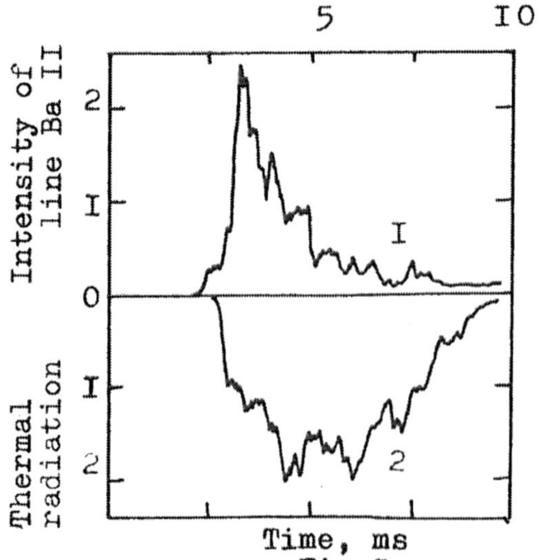

Fig.I

At the initiation of the discharge the radiation of Ba II 455,4 nm spectral line is a structure shown in fig.2. The delay between the switching-on of the voltage on the lamp and appearance of Ba ions radiation is 80-200 mks.

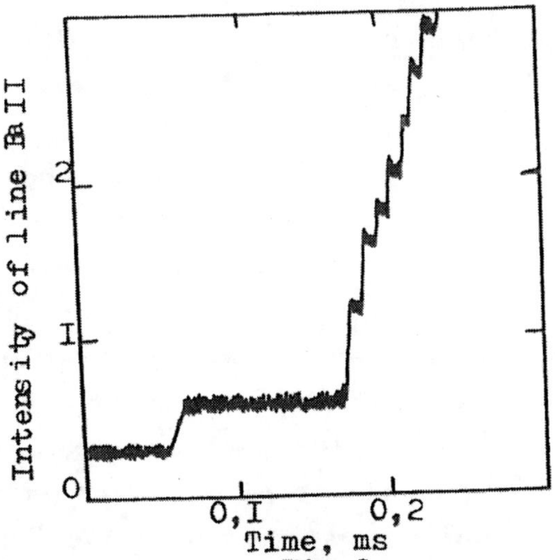

Fig.2

Apparently, the structure of the radiation (fig.2) is connected with the passing and gradual increasing of the electron avalanches accompaning the breakdown of the gas discharge interval.

The analysis shows that the Ba ion emission and heat radiation appear 300 - 500 mks earlier than the electric arc with cool cathode is formed and the decrease of the voltage drop on the lamp takes place. This is illustrated by fig.3, where curves I and 2 correspond to ion emission and voltage on the lamp.

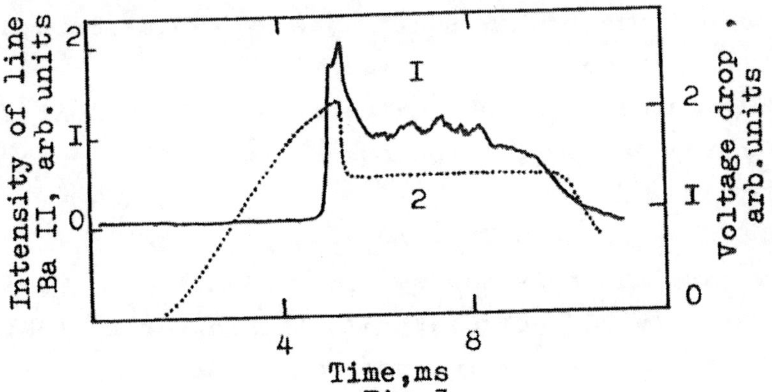

Fig.3

The obtained results show that the components of the emissive coating to a great extent determine the kinetics and conditions of the breakdown of the gas discharge interval, electric field strength as well as the mechanizm of electronic emission of the cathode spot.

n° 29

Colorimetric efficiency index as the new quantity parameter of the lamps

Wojciech S T A N I O C H, Dr eng.
Centralny Ośrodek Badawczo - Rozwojowy POLAM, Warsaw, Poland

The method to be tested nowadays to appreciation of the lamp quality involved with the spectral power distribution are: luminous efficiency, correlated colour temperature and colour rendering index. The values of the luminous efficiency and CRI for the given CCT are contradictory one to another on account of two different criterion of appreciation. Luminous efficiency is only quantity comparison parameter of the light sources especially of the same types, however it is not suitable criterion for the lighting engineering. The best appreciation method of the quality as well as quantity lamps parameter from point of view of the lighting engineering would be the visibility level /VL/ involved with the SPD. Such parameter is the colorimetric efficiency index /CEI/ determined on the basis of luminance and chromaticity contrast of the standard visual task. Values of CEI indices are in good agreement with CRI indices, because the both parameters take into account the psychophysical characteristics of the observer, that is colour as well as light vision, for which, of course, it is essential the full spectral visual range of radiation but not only maximum of spectral luminous efficiency.

A general analytic model of CEI calculation is the following:

1/ calculate the saturation indices SI according to formula

$$/SI/^2 = /u_i - u_o/^2 + /v_i - v_o/^2$$

in which u_i v_i are the tristimulus coefficients of 8 Munsell samples /the same as to CRI calculation/ illuminated by tested or standard light sources /illuminents/ of the same correlated colour temperature and u_o v_o are the tristimulus coefficients of the tested or standatd light sources

2/ calculate the mean for 8 samples equivalent visibility level

$$VL_{eq} = /0,285 + 0,715 /SI^2 : SI_r^2 //^{1/2}$$

in which SI is the mean for 8 samples saturation index for lamp to be tested and SI_r is the mean for 8 samples saturation index for illuminant of the same correlated colour temperature

3/ calculate the colorimetric efficiency index CEI for adaptation luminance 1 cd/m^2 and 30 cd/m^2

$$CEI_1 = 1,64 / 2,22\ VL_{eq}^{-0,4} - 1 /^{-2,5}$$

$$CEI_{30} = 0,0546 / 1,313\ VL_{eq}^{-0,4} - 1 /^{-2,5}$$

In the following table there are given as the example the numerical values of CEI for typical light sources

Light sources	CEI_1	CEI_{30}
High pressure sodium	0,72	0,50
High pressure mercury	0,95	0,88
Fluorescent standard	0,90	0,80
Fluorescent de luxe	1,00	1,00

Corrected values of luminous efficiency are calculated as the product $\eta_{corrected} = \eta \cdot CEI$

Values of CEI calculated by means of analytic method were also examined by means of direct psychophysical tests

n° 30

LIGHT SOURCES INVESTIGATION BY MEANS OF COORDINATE GAS COUNTERS.

G.F.Karabadjak, V.D.Peskov
Institute for Physical Problems
ul.Kosygina 2, Moscow USSR

Gas discharge light sources elaboration is impossible without detailed study of processes happening there. For investigation of processes in lamps and optimisation of their parameters coordinate gas counters (CGC) are elaborated and utilized at the Institute for Physical Problems of the USSR Academy of Sciences[1]. Such devices allow the radiation with $\lambda = 100 \div 2000$ °A to be detected with the spectral resolution $\Delta\lambda = 1 \div 50$ °A and spatial resolution ~0,1 mm. A spatial distribution of the absorption coefficient and absolute intensity of radiation may be also measured by using CGC. One can utilize CGC as a detector of radiation in the focal plane of a spectrograph.

A glow discharge at the pressure $p = 10^{-3} - 10^{3}$ Torr, electrode and electrodeless HF and UHF discharges ($p = 10^{-1} - 2{\cdot}10^{4}$ Torr) [2], a DC arc discharge ($p = 10^{-4} - 2{\cdot}10^{4}$ Torr) [2], [3], the compact excimer radiation source based on the radioactive preparation ^{55}Fe [4], the VUV source created on the basis of high current electron gun [5] were investigated with the help of CGC. Such investigations permitted us to get essentually new information about physical processes in light sources. For example, we managed to observe HF and UHF discharges while breaking down into filiforms, to discover the instability related to the excited atoms accumulation followed by the gas breakdown, to study the electrode region structure of discharge. The CGC advantages become especially distinct when investigating high pressure discharges, since the radiation from the cold outlying region masks the phenomena in the more hot central part of discharge.

It is of great importance that CGC allow not only spectral but the other physical measurements to be carried out. Atomic temperature radial distribution from X-ray absorbtion measurements [5], electrons work of exit from lamp electrodes measurements, determination of the probability of electrons exit from the cathode under the positive ions bombardment γ_{+} etc. are the examples. One of the examples - the measurements of γ_{+} coefficient in glow discharge is illustrated below.

A glow discharge was lit up in the coordinate cylindrical counter with the multisectional cathode. So it became possible to measure γ_{+} in different points of the cathode. A cathode spot was formed on the one of the cathode sections. The anode and each of the cathode sections were connected to their own preamplifiers, so that the collected charges and the currents flowing through the anode and each of the cathode sections were controlled independently one from the other. Experiments were carried out in air, methane, noble gases and in noble and molecular gases mixtures ($p = 10^{-4} - 10^{3}$ Torr).

A stable glow discharge was rapidly ($t \leq 0.1 \mu s$) turned off, right after its appearance. When the major part of positive ions was collected on the cathode sections ($t \sim 10 \mu s$, $p \sim 10$ Torr) and current stopped, the collimated impulsive VUV radiation was brought into the counter volume after the fixed time delay. This radiation ionized the gas, and quite definite number of ions 10^6 occured in the counter volume. Electrons were getting together on the anode after $\sim 0{,}1 \mu s$ ($p \sim 10$ Torr) and ions after some tens of microseconds - on the cathode. The secondary electrons, occuring while the ions were neutralized on the cathode assembled during $\sim 1 \mu s$ on the anode, which resulted in the secondary impulse of voltage there. By the correlation of the amplitude of this secondary impulse with the VUV radiation intensity, it was possible to find out γ_+ in the predetermined moments. The electrons work of exit was estimated in the same way. These measurements showed that the γ_+ was a few times greater on the sector where cathode spot was located, compared with the neighbouring regions, and after glow discharge was turned off γ_+ was at first decreasing rapidly (time constant $\sim 10 \mu s$) and then slowly (time constant $\sim 10^{-2} s$).

It is known that the ion bombardment cleans partially the electrode surface. On the other hand, the less absorbed gas monolayers are held on the electrodes surface, the more γ_+ is. So, it is natural to attribute the rapid component to the surface cleaning under ions bombardment.

Given examples show that CGC-methods permit to investigate in detail the gas discharge light sources, which is of greate importance for their practical design.

References:

1. V.D.Peskov 15- ICPIG 1735, Minsk 1981
2. V.D.Peskov Journ. de Physiq. 40, 333-336, 1979
3. G.F.Karabadzak, E.L.Kosarev, V.D.Peskov, E.R.Podoljak. Tezisi dokladov Vsesojuznoj Konferencii VUF-82, 231, M.1982.
4. A.V.Mitrofanov, V.D.Peskov. Tezisi dokladov Vsesojuznogo seminara VUF-78 64, Leningrad 1978.
5. V.D.Peskov. G.P.Prudkovskij. Tez. dokl. Vses.Konf. VUF-82,277 M.1982

n° 31

Ball Lightning - A Side Step of Light Research

E. Fischer
Philips GmbH Forschungslaboratorium Aachen, Germany

Ball lightning is one of the few phenomena on earth that have not found a satisfactory explanation by the physical sciences up to now. The rare occurrence and the inability to produce ball lightning in the laboratory has led many scientists to deny the existence of ball lightning at all. But from the large number of eye-witnesses it must be concluded, that there is some real luminous spherical phenomenon in nature, to which at least part of the reported observations refer.

Obviously the existence of some unexplained light emitting object should stimulate light researchers to check, whether the observed effect may be used in an artificial light source, as nearly all other light emitting processes in nature have been applied in some form - beginning from the discharge in a lightning flash over various kinds of flames to the mild shine of glow-worms or fluorescing bacteria.

From the eye-witness descriptions some general properties of ball lightning can be deduced, though there is a large scatter in the reported data. Most frequently the diameter is given as 10 to 50 cm, the colours range from bluish-white to yellow red, the estimated life times are in the order of 1 to 10 seconds. The phenomenon is terminated either by quiet extinction or by abrupt explosion.

Various explanations have been proposed for the ball lightning phenomenon, some as exotic as matter-antimatter annihilation. But regarding the fact that ball lightning is mostly associated with normal thunderstorms, only those explanations should be discussed seriously, which are based on states of matter that might occur in thunder-storms. Thus only electrical discharges or chemical processes can be responsible for the energy production in the ball. Theoretical models involving a steady supply of electrical energy should be ruled out, however, as well as explanations that interpret the ball as a hot volume of gas, because in that case magnetic or buoyancy forces would destroy any spherical configuration within tenths of a second.

The combustion of an aerosol, however, which has about the same density as the surrounding cold air, fulfils all requirements for a stationary spherical luminous phenomenon. Such aerosols can be formed, when an ordinary lightning stroke hits an organic piece of matter (e.g. a flying bird). The material is heated adiabatically until the increasing vapour pressure of the enclosed water disrupts it into small particles, which are then heated to a high temperature by the energy of the lightning flash. After its decay the combustion continues in the surrounding air. Quantitative estimations of the energy balance of such a combustion process show that the burning temperature, the life-time and even some more detailed properties of the model meet the observed data.

In conclusion the explanation of the ball lightning phenomenon has not provided us with a new kind of light source, but it shows that the mysterious fire balls are nothing but a special form of those combustion processes which we observe in nature every day.

PHENOMENA OCCURING IN FLASHTUBES

M.SKOWRONEK Y.VITEL
E.R. PLASMAS DENSES
UNIVERSITY P. ET M. CURIE
75230 PARIS (FRANCE)

1-Introduction

Flashtubes are well-known devices used in numerous fields owing to their properties:high brightness level,short time illumination ,spectra containing a strong continuum,high ultra-violet output,high repetition rate,possibility to choose the spectral distribution,and long lifetime.The industrial applications of the flashtubes are ,for instance:airport or highway signalisation,high speed photography,laser pumping,drying of inks or dyes.

Plasmas created in flashtubes have an elctron density about $10^{18}cm^{-3}$ and an elctron temperature of 1ev which place them at the limit between the classical kinetic plasmas and the strong correlated plasmas. Consequently,the flashtubes have a great interest for the study of non ideal plasmas effects on the transport coefficients and on the light emission.In this paper,we will describe the different phases of the plasma development in flashtubes and their influence on the light output.

2-Experimental set-up

We study linear flashtubes which are cylinders of 15cm interelectrode spacing and 10 or 6mm inner diameter.They are filled with noble gases under pressures between 200 to 600 torrs.They are designed by the society "Verre et Quartz".The electrical feed line is made of five LC
cells performing a good impedance matching between the line and the plasma.This pulse forming network gives a trapezoidal current pulse with a 1ms plateau constant whithin 1%.The intensity and the voltage across the tube are precisely measured with a precision of 1%.The light emtted is analyzed by a high speed camera,by a monochromator followed by aphoto-multiplier,and also a C.C.D. device

3-Study of the plasma expansion in the flshtubes

Different phases in the plasma expansion are observed:

a) A time lag between the triggering and the development of the discha -rge

b) A shock wawe starts simultaneously with the channel growing

c) A fast expansion phase.The plasma column grows rapidly before it stops because its interaction with the reflected shock.

d) Aslow expansion phase.The expansion is now acompanied with oscillations of the column boundaries.After the damping of these oscillations,we can see at few places of the column,1cm between them,contractions of the plasma column.These local contractions appear and develop more or less depending on the gas nature,the initial pressure and the energy of the discharge.

The study of the different phases allows us to determine an effective diameter of the discharge and to measure the plama parameters when these perturbations are damped or of a low level.

4-Measurements of the absolute spectral brightness and determination of the plasma parameters

In the part of the spectrum where the plasma may be considered as optically thin,we have measured the spectral transverse ditribution and we have deduced the radial distribution by inversion of the Abel integral equation.The emission coefficient is proportionnal to $N_e^2/\sqrt{T}$. Applying Saha equation we have determined the values of $N_e(r)$ and $T(r)$.

5-Study of the Stark broadening of noble gases lines(ArI,ArII, KrI,KrII,XeI,XeII,)

We have recorded few lines and we have measured their broadening, adjusting the line profile to a Lorentzian.Some of the broadening values normalized to $N_e=10^{17} cm^{-3}$ are in good agreement with other people's measurements.These values are not in agreement with available theories.We have also measured the shift Of these lines.

6- Conclusion

Flashtube studies allow the determination of the conditions leading to a good plasma development and achievement of brightness unformity which is essential in some applications.This lamp may also serve to the measurement of dense non-ideal plasma properties.

n° 33

STUDY OF A DISCHARGE SOURCE FOR VACUUM U.V. ABSORPTION TECHNIQUE.

M. MORLAIS*, J.L. QUEFFELEC** and J.C. GOMET**
Laboratoire de Spectroscopie, ERA C.N.R.S. 748* and Département de Physique Atomique et Moléculaire**, Université de RENNES I, 35042 RENNES CEDEX, France.

B.R. ROWE,
Laboratoire d'Aérothermique du C.N.R.S., 4 ter route des Gardes, 92190 MEUDON, France.

Y. GUERN,
Laboratoire de Spectroscopie Quantique, Université de Bretagne Occidentale, 29283 BREST CEDEX, France.

Absorption technique can be very powerful for diagnostic of gases and plasmas and have been widely used. This technique requires to have a source of radiations, the absorption of which can yield the density of atoms or molecules in a given state in the studied gas |1|. In the case of atoms or atomic ions, the wavelengths used for this purpose lie very often in V.U.V.. The main problem is that the profiles, or at least the half-widths, of the lines emitted by the source, are needed in order to obtain a correct value for the density. On an other hand, if one knows the density, this technique gives indications on the broadening of the line.

In our laboratories, a vacuum U.V. source of radiation of penning kind (low pressure discharge) |2| |3| was studied for the case of the 919.78 Å and 932.05 Å lines of Ar II and the 1200, 1493 and 1744 Å multiplets of NI.

In the case of Ar II, the total density of Ar^+ ions was determined by a Langmuir probe technique |4| and the results are in very good agreement with those obtained by the absorption technique if one assumes that the Ar II profile lines are doppler like at the source temperature. This is not the case for NI lines, absorption procedure gives indication of a strong broadening of the lines, while the calculation of the usual causes of broadening does not indicate such an effect.

A prototype of Fabry-Perot interferometer |5| was then used to study the 1494.67 Å and the 1742 Å lines of NI. Despite the very difficult problems linked with vacuum U.V. interferometer, this prototype has a resolution better than usual spectrometer ; however it cannot be used up to now below 1400 Å.

It was found that the lines are strongly broadened if the discharge source is used with pure nitrogen. The broadening is lower if an helium buffer is used with a small amount of nitrogen ; then keeping a constant partial pressure of nitrogen, an increase of the helium pressure reduces strongly the broadening. These results are shown in fig. 1 for the 1494.67 Å line.

The absorption procedure shows that this is also the case for the others NI lines, in particular for the 1200 Å multiplet. Fig. 2 displays the absorption of these lines, with exactly the same absorption medium, with or without a buffer gas.

It can be shown that auto-absorption in the source cannot explain such large effects, so our results strongly suggest a specific mechanism of broadening in the case of nitrogen.

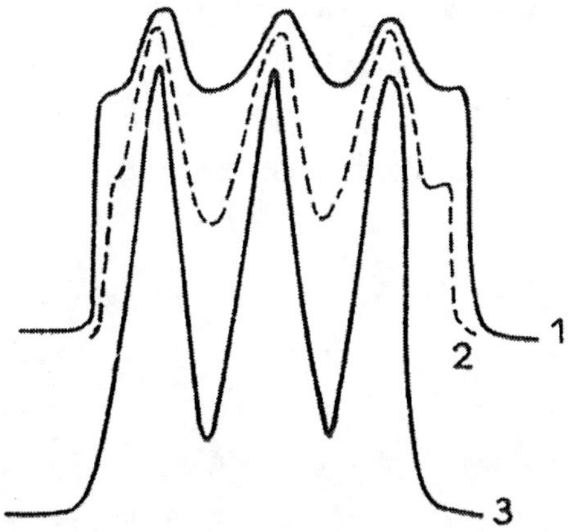

Fig. 1 - 1. N_2 (0,4 torr).
2, 3 - He buffer, constant pressure of N_2 total pressure (torr).
p_2 = 0,8 ; p_3 = 2.

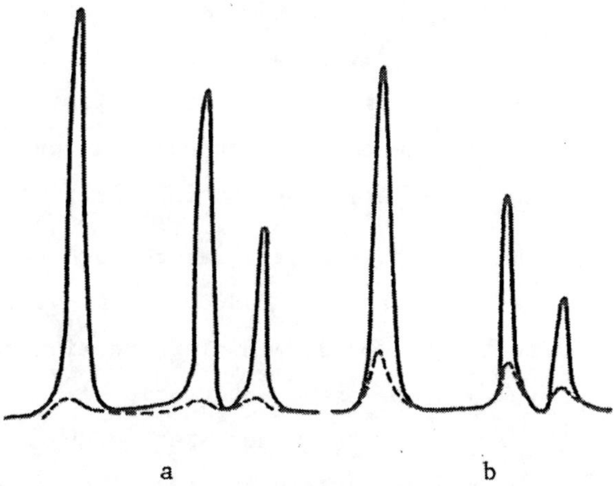

Fig. 2

a - Buffer He   b - N_2 pure

REFERENCES.

|1| MITCHELL A.G. and ZEMANSKY M.W., 1981 *Resonance radiation and excited atome* Cambridge Univ. Press, N.Y.

|2| DAMANY H. and BOURSEY E. *(Private communication)*.

|3| GERNOGORA G., 16 December 1980, *Thèse Doctorat,* Université de Paris-Sud.

|4| DUDECK M. et al (to be published).

|5| BIDEAU-MEHU A., 1982, *Thèse Doctorat,* Université de Brest.

SPECTRAL STUDIES IN Na-Sc METAL HALIDE LAMPS

Harold L. Rothwell
William M. Keeffe
HID Development & Research Laboratory
Sylvania Lighting Center
Danvers, Mass. 01923

Owing to its good color and high luminous efficacy the Na-Sc doped high pressure mercury arc has had good acceptance for applications of general lighting. For the most part, its development as a light source has evaded rigorous theoretical treatment primarily because of the complexity of a multi-additive system. In this presentation, we will describe spectroscopic approaches which provide insight into the composition characteristics of a metal additive arc.

The development of a method to determine the concentrations of metal additives in the arc core has extended over many years. In 1975, at this conference, we presented a technique for determining additive partial pressures in the plasma which relied on the line-of-sight integration of optically thin emissions. Since that time much effort has been undertaken to develop a systematic experimental approach to make the measurements rapid and reliable. The culmination of these efforts is a fast sampling polychromator which was detailed at the 1980 Illuminating Engineering Society (IES) Conference.[1] Subsequent work on vaporization rates employing the characteristics of the polychromator were reported at the 1981 IES Conference.[2]

In the last few years, the technology of lamp development has been extended to include highly loaded high pressure (~10 atm) arc sources that provide high luminous output (~ 100 LPW) with low source color temperatures (~3000°K). These types of sources have proven to be unacceptable for line-of-sight techniques because of the difficulty of finding interference free optically thin emission lines. In order to extend our basic knowledge of the plasma and derive information about the additive levels in the arc, we have developed a spectral technique which is based on the pressure broadening of resonance line radiation. The procedure, similar to that described by DeGroot[3] and Chien Yu-Min[4] for HPS lamps, adapts the impact approximation to the optically thick self-reversed resonance lines of Na, Sc and Hg. Both resonance and van der Waals interactions have been considered. Unlike HPS lamps, the presence of molecular species e.g., NaI and ScI_3, require chemical equilibrium calculations to determine the depletion of atomic species away from the core of the plasma. The analysis procedure has been reduced to practice and a summary of results for a typical metal halide lamp is shown in Figure 1.

We have tested our method with several experiments designed to provide basic information on the spatial and temporal behavior of the metal additives in a vertically operated alternating current arc.

The test lamp was constructed from tubular fused silica with contoured ends in which tungsten rod electrodes were sealed. By analyzing the self-reversed resonance lines of Na (589.0, 589.6), Sc (390.7, 391.1), and Hg (253.6) coupled with detailed plasma temperature profiles, we measured the axial segregation of Na and Sc at various phases of the AC cycle. Detailed mercury pressure measurements at several axial positions have verified that the Hg pressure commutes and that a 10% modulation in Hg pressure occurs between peak current and zero crossing times. Interestingly the levels of Na and Sc show no appreciable modulation with phase.

The consistency of the experimental results has provided confidence in the model used for the analysis of line shapes. We plan to improve our model by further refinements in the treatment of the interaction potentials and the radial temperature profiles.

References:

1. H. L. Rothwell and W. M. Keeffe, Journal of IES, 10, 1980.
2. H. L. Rothwell and W. M. Keeffe, Lighting Design and Application, January, 1982.
3. de Groot J.J., "Investigation of the High Pressure Sodium and Mercury/Tin Iodide Arc" 1974 (unpublished).
4. Chien Yu-Min, J. Appl. Phys., 51, 1980.

OPTICALLY THIN			OPTICALLY THICK			
			RESONANCE			Van Der Waals
Na(616.1)	Sc(563.1)		Na(589.0)	Sc(390.7)	Hg(253.7)	Na(589.0)
P_{na} (Torr)	P_{sc} (Torr)	P_{Hg}* (Torr)	P_{na} (Torr)	P_{sc} (Torr)	P_{Hg} (Torr)	P_{Hg} (Torr)
2.9	1.2	2700	4.0	2.1	2510	1940

*CALCULATED FROM ARC TUBE DIMENSIONS AND MERCURY FILL

PARTIAL PRESSURES OF ADDITIVES DETERMINED FORM LINE INTENSITY MEASUREMENTS

n° 35

Modelling of High Pressure Gas Discharges Including Radiative Transport

H.-P. Stormberg
Philips GmbH Forschungslaboratorium Aachen, Germany

A model for cylindrical high pressure gas discharges has been developed which requires only some simple lamp data (diameter, length, mercury charge) and data forthe lamp circuit (circuit equation, applied voltage U(t)) as input parameters. It calculates the time dependence of the radial temperature profile T(r,t), pressurep(t), lamp voltage U(t) including the re-ignition peak, lamp current I(t), wall temperature of the arc vessel and the luminous flux. Thus it can be used for calculating the properties of high pressure lamps operated from conventional or electronic power supplies. Only the value of the electrode voltage drop cannot be calculated and must be estimated or determined experimentally. As the voltage drop does not vary significantly for different high pressure lamps, an estimated value (about 20 V) should always be a sufficient approximation. The influence of axial convection has not been included in our model.

One of the main difficulties when evaluating such models is the determination of the so-called local net emission of the plasma column (cf. Ref. 1). The calculation of this radiation term is extremly complicated because a contribution due to self-absorption is involved which cannot be expressed as a function of local plasma parameters (temperature, particle number densities) but depends on the whole temperature profile. The procedure for the solution of the problem runs as follows: The broadening constants of all relevant optically thick mercury and sodium lines have been determined experimentally. With these constants the local net emission is obtained for an arbitrary temperature profile by solving the radiation transport equation. The numerical calculation of the local net emission, which is very time consuming is replaced by an analytical approximation which is valid over a broad range of lamp diameters and mercury pressures. The same approach has been applied to calculate the light flux and the energy flux transmitted through the quartz vessel. Using these results the luminous efficacy and the wall temperature have been derived from the model, too.

The applicability of the model including the validity of the approximations is demonstrated for several high pressure mercury discharge conditions. In a series of dc experiments with varying current, lamp diameter and mercury pressures the temperature profiles, wall temperatures and the luminous fluxes have been measured. The results are compared with the respective data evaluated from the model. In further experiments the time dependence of lamp voltage, current and temperature profile has been measured for sinusoidal and pulsed supply voltages. The corresponding model calculations are in good agreement with the experiment. It becomes evident that the local net emission cannot be expressed in terms of the local temperature only, but is governed by the shape of the whole temperature profile which changes strongly during an ac cycle.

/1/ R.J. Zollweg, J. App. Phys. 49, 1077 (1978)

n° 36

EFFECT OF THE TEMPERATURE AND DENSITIES MODULATIONS ON THE SHAPE OF SELF REVERSED VISIBLE TRIPLET LINES IN HIGH PRESSURE DISCHARGES OPERATED UNDER A.C. CONDITIONS

Jean-Jacques DAMELINCOURT, Michel AUBES and Dominique GUILHEM
Centre de Physique Atomique, Laboratoire Associé au CNRS n° 277, Université Paul Sabatier, 118 route de Narbonne, 31062 TOULOUSE CEDEX, France
Dimitrios KARABOURNIOTIS
Department of Physics, University of Crete, IRAKLION, Greece

In sources operated at a.c. power lines frequencies the light output is modulated in response to the variation of instantaneous power input.

The shape of observed lines profiles depends on the radial variation of the properties (temperature and concentrations of different species) of the radiating layer and on a number of constants that characterize the dependence of the line broadening on the physical conditions in the plasma.

In this work an attempt is made to analyse the effect of the variation of the plasma properties and the experimental uncertainties and model inaccuracies on the characteristic parameters of the calculated and measured self-reversed lines.

The temperature distribution over the radius of the source was found from side-on mesurements of the absolute intensity of the mercury line 577 nm corrected for low absorption using the Abel inversion technique. The iteration procedure also gives the instantaneous pressure in the arc.

We have performed calculation of the line profile shape by solving the integral form of the radiation transfer equation.

The impact approximation is used for the description of the broadening due to electrons and ions. The electrons move rapidly through the gas and the assumption of impact appoximation is fulfiled for the electrons even in the far part of the line wings. For ions this assumption remains no longer valid in the red line wing but their contribution to the total broadening is not predominant for the discharges studied. For collisions between neutral atoms we have considered a Van der Waals interaction potential.

For these collisions impact approximation may describe the central part of the line. The broadening of the red wing due to atoms must be described by a Van der Waals interaction according to the quasi-static approximation. We have also considered the contribution of the Doppler effect. This quasi-static effect in the wing is accounted for according to the calculations by Stormberg (2). A more detailed description of the calculations is given in reference (1).

The calculation of the line shapes was carried out with variation of the parameters which characterize both the model of the plasma and the properties of the spectral lines.

From the analysis of the calculated and of the experimental data one may conclude

a) Only the inclusion of the quasi-static broadening allow the description of the asymetry in the red wing of the self-reserved line.

b) Uncertainty in the knowledge of broadening by electrons does not appreciably change the calculated line shape at the moment of minimum emission.

c) Uncertainty in the value of the transition probability for the 577 nm line does not seriously affect the determination of C_6.

d) The distance between maxima of the self reversed line does not depends appreciably on temperature.

From these remarks it is seen that from comparison between theoretical and experimental results, using the variation in the electronic density at 5 and 0 ms, it is possible to evaluate a set of parameters which would be primary valid for the case of self reversed lines.

Some of the results obtained are given in table 1.

TABLE 1 - Experimental values of C_6 and A.

line nm	404.7	435.8	546.1
$C_6(10^{-30}$ rad cm^6 $s^{-1})$	1.5	1.5	1.6
$A(10^6$ $s^{-1})$	18	55	54

REFERENCES

1. Internal report.

2. H. P. STORMBERG, J. Appl. Phys. 51, 1963 (1980).

Laser scattering from metal halide arcs

Th. Zaengel and E. Groiss
Philips Research Labs Aachen, Germany, and Eindhoven, The Netherlands.

Electronic Raman scattering (RMS) and near resonant Rayleigh scattering (RLS) have been employed to determine absolute local densities of groundstate and metastable atoms in a Na-In-Tl-I-Hg arc. Because of the high selectivity (inherent to RMS, obtained by resonance enhanced cross sections in the case of RLS) and the direct scattering mechanisms involved, the 90 degrees scattered signal can be immediately converted into local densities, after calibration of the experimental set-up with RLS in air or nitrogen. No information about the composition or state of the plasma is used.

We reported earlier on RMS on groundstate In and metastable Tl atoms using Kr^+-laser radiation*. Since the Tl groundstate was not accessible with this laser, we now employed an UV-Ar^+-laser, which has suitable lines to measure both the groundstate $^2P_{1/2}$ (363.8 nm → 507.8 nm, $\sigma_{zz} = 2.1 \cdot 10^{-29}$ m^2, $\rho = 1.09$) and metastable $^2P_{3/2}$ state (528.7 nm → 374.4 nm, $\sigma_{zz} = 7.9 \cdot 10^{-28}$ m^2, $\rho = 1.01$). The figure gives the (time-averaged) radial profiles of both states in a 400W AC operated arc and the resulting distributional temperature. The value for the wall is obtained from its IR-emission. Good agreement with averaged temperatures from emission measurements is obtained on the axis. With regard to investigations into a possible non-equilibrium situation in the outer arc region, exact temperature measurements for comparisons are lacking.

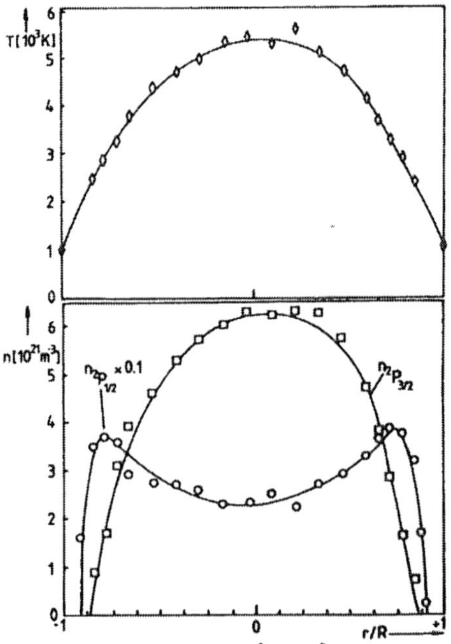

Radial density profile of Tl $^2P_{3/2}$ and $^2P_{1/2}$ in a Na-Tl-In-I-Hg arc and resulting temperature

In order to determine Na groundstate $^2S_{1/2}$ densities we coupled the Ar^+-laser to a ring-laser, which allowed us (with R6G) to tune into the blue wing of the Na D-line 589.0 nm. The RLS cross section at 0.6 nm from the line center, a typical distance in our experiments, is $1.38 \cdot 10^{-24}$ m^2 ($\rho = 0.04$)**, yielding ample selectivity against other plasma components. Nonetheless the beam attenuation is less than 5 %. The fact that we have almost completely polarized scattering, allows to compensate for straylight and multiple scattering on the quartz wall by rotating the polarization of the incident beam and substracting the signal thus obtained. This simplifies the (necessary) correction considerably. We have been able to measure up to relative radial coordinates of .8 to .9, where the straylight level passes quickly the real signal.

The high cross section and resulting comparably stable signals allow for a scan of a 50 Hz AC cycle using box-car techniques within a couple of minutes. The figure shows an example of such a measurement at the axis of a vertically operated 400 W arc (in the middle between both electrodes, straylight subtracted). In the current maxima sodium is highly ionized, as indicated by the two minima in the groundstate (~ total atom) density. Strong ion transport in the electrical field occurs, which becomes visible in the following temperature minima, where most of the ions recombine. Since the residual ionization does hardly differ between the two minima (about the same temperatures from emission spectroscopy), the relative heights of the two density maxima give a direct impression of the changes in the sodium concentrations at given position during the AC cycle in these arcs with strong axial segregation.

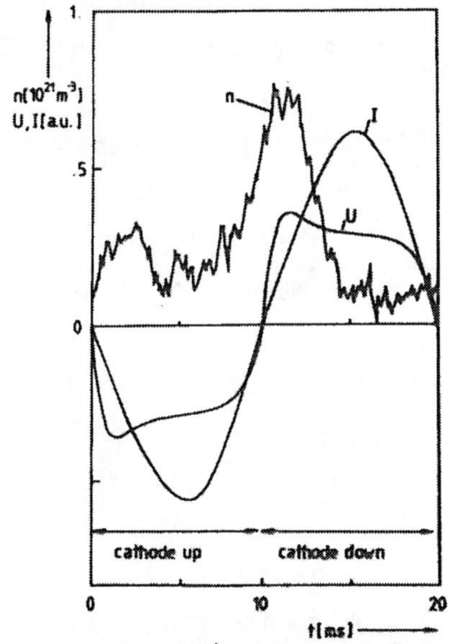

Variation of $Na\,^2S_{1/2}$ density in a Na-Tl-In-I-Hg arc within the AC cycle

In order to study the influence of the oscillating ion fluxes on the axial segregation we measured time averaged densities for three axial positions on the axis of the arc as a function of frequency (see fig.). The arc is square wave operated in order to obtain stable arc temperatures (and fractional ionization) only influenced by the density changes themselves. The power is kept constant to 400 W, and so were current and voltage, since the U/I characteristic was obviously hardly influenced. Nonetheless we observe clear changes in the axial segregation and the densities, especially in the upper parts of the arc. Since the ion "swing" in the external oscillating field approaches zero towards higher frequencies, the high frequency side can be taken as a reference situation in which the effect of the external field cancels itself locally, that is to say, the particle distributions are governed by the interaction of convection and the various internal diffusion processes. Going from higher frequencies towards lower ones, one sees the increasing interference of the ion oscillation caused by the external field with the forementioned processes, resulting in changed equilibria.

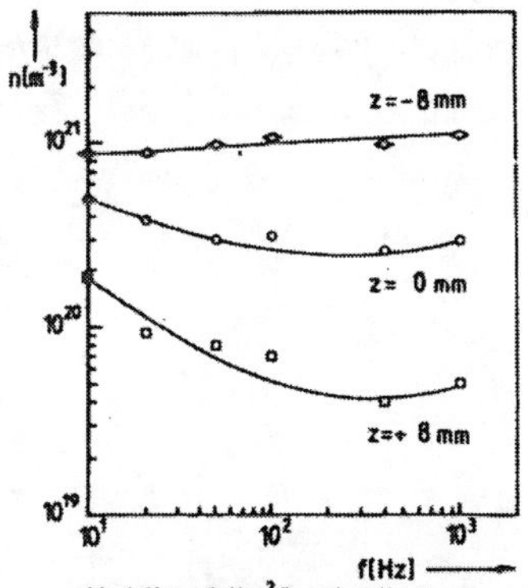

Variation of $Na\,^2S_{1/2}$ density in a Na-Tl-In-I-Hg arc with frequency for three axial positons

* Bull. Am. Phys. Soc. 25 (1980) 99

** Appl. Phys. 11 (1976) 253-7

n° 38

THE APPLICATION OF NON-LINEAR LASER SPECTROSCOPIC TECHNIQUES TO THE INVESTIGATIONS OF OPERATING LIGHT SOURCES

R. DEVONSHIRE, Department of Chemistry, University of Sheffield, SHEFFIELD, U.K.

There are relatively few reports in the literature of the application of the techniques of conventional linear absorption and emission spectroscopy to the study of light sources. There are clearly strong practical constraints such as self-luminosity and optical inhomogeneity which, at least in part, account for this observation but equally clear is the requirement for, and the enormous potential value of, a non-perturbing diagnostic technique that can be applied to operating light sources.

In recent years a number of new non-linear laser spectroscopic techniques have reached a stage of development where we believe they are particularly relevant to the science and technology of light sources. The techniques are successfully being applied to systems whose operating conditions are just as vigorous and complex as those in regenerative chemical cycle incandescent and in discharge type light sources.

In the talk the various techniques will be briefly described and the scope of the results to be obtained and the limitations on their interpretation will be discussed. The measurement of the physical characteristics of the environment (temperature and species partial pressures) within the envelopes of operating light sources and of the chemical species present during operation will be discussed and, hopefully, new results will be presented.

n° 39

INFLUENCE OF ASYMMETRIC SELF-REVERSAL OF SPECTRAL LINES ON THE PLASMA TEMPERATURE DETERMINATION.

D.Karabourniotis, M. Drakakis and C. Karras
Department of Physics, University of Crete, Iraklion, Crete, Greece
J.J. Damelincourt
Centre de Physique Atomique, Laboratoire Associe au CNRS,Universite P.Sabatier, Toulouse, France

Recent papers[1-3] suggest the application of the Cowan and Dieke theory[4] of symmetrically self-reversed spectral lines for diagnostic purposes in axisymmetric arc discharges,which are in local thermodynamic equilibrium. If the plasma is highly inhomogeneous then many of the lines are asymmetrically self-reversed. An attempt is made here to determine the effect of the line asymmetry on the temperature determination,by connecting the asymmetry with the relative displacement of the line center. For that we consider a local line profile described by a dispersion function,

$$P(w,y)=(1/\pi\delta_0)P'(w,y),$$

with $$P'(w,y) = \delta'(y)\{[w-\eta\delta'(y)]^2+[\delta'(y)]^2\}^{-1},$$

in which $$w = (\nu-\nu_\eta)/\delta_0, \qquad \delta'(y)=\delta(y)/\delta_0=\exp[-b(1-y)^2],$$

$\eta=\Delta(y)/\delta(y)$, $$y = \int_x^{+x_0} \bar{n}_a(x')dx' \text{ and } \bar{n}_a(x)=n_a(x)/N_a.$$

$n_a(x)$ is the density of absorbing atoms; N_a is the number of absorbing atoms along the line of sight within the plasma boundaries $-x_0$ and $+x_0$; δ_0 is the half-width at the arc axis; b is a parameter which characterizes the radial course of the half-width $\delta(y)$ whithin the emitting zone with $b \geq 0$; $\Delta(y)$ is the line shift;and ν_0 the central frequency of the undisplaced line. It is assumed that the parameter η is constant throughout the plasma. Then,the shape of the observed line may be written in the form

$$I(w) = W_{\nu_o}(T_m)pJ(W) \tag{1}$$

Here $W_{\nu_o}(T_m)$ is the Wien's function for the temperature T_m (maximum temperature along the line of sight), p is the absorption parameter and

$$J(w) = \int_0^1 y^{n-1}P'(w,y)\exp(-p\int_0^y P'(w,y')dy')dy')\{1+\exp[2p(\int_0^y P'(w,y')dy'-\int_0^1 P'(w,y')dy')]\}dy$$

is the relative intensity,where n is the inhomogeneity parameter.

From Eq.(1) the temperature is given by

$$T_m=(h\nu_o/k)[\ln(2h\nu_o^3/c^2)+\ln K_{b,\eta}-\ln I_1]^{-1} \tag{2}$$

where I_1 is the measured absolute intensity at the larger maximum of the line and $K_{b,\eta}=pJ_1$ is independent of the value of p. Fig.1 gives $K_{b,\eta}$ as a function of n for various b and η. For the given source and the given line we determine the values of n,p and b we calculate the relation $J_1/J_2=f(\eta)$ (Fig. 2). From this figure and the experimentally measured ratio I_1/I_2, where I_2 is the smaller maximum of the line, we determine η. So, if one knows n, b and η, it is possible to find the value of $K_{b,n}$ and, from Eq.(2), the temperature T_m by taking into account the line asymmetry I_1/I_2 An application is made using the Hg 5461 Å line emitted by a 3 atm mercury discharge operated vertically at 50 Hz, 100 W/cm (Table)

n	p	b	η	$K_{o,o}$	$K_{b,o}$	$K_{b,\eta}$	$T_m(0,0)$	$T_m(b,0)$	$T_m(b,\eta)$	T_m(Bartels)
1.33	30	0.54	0.925	0.664	0.680	0.695	6350	6320	6290	6150

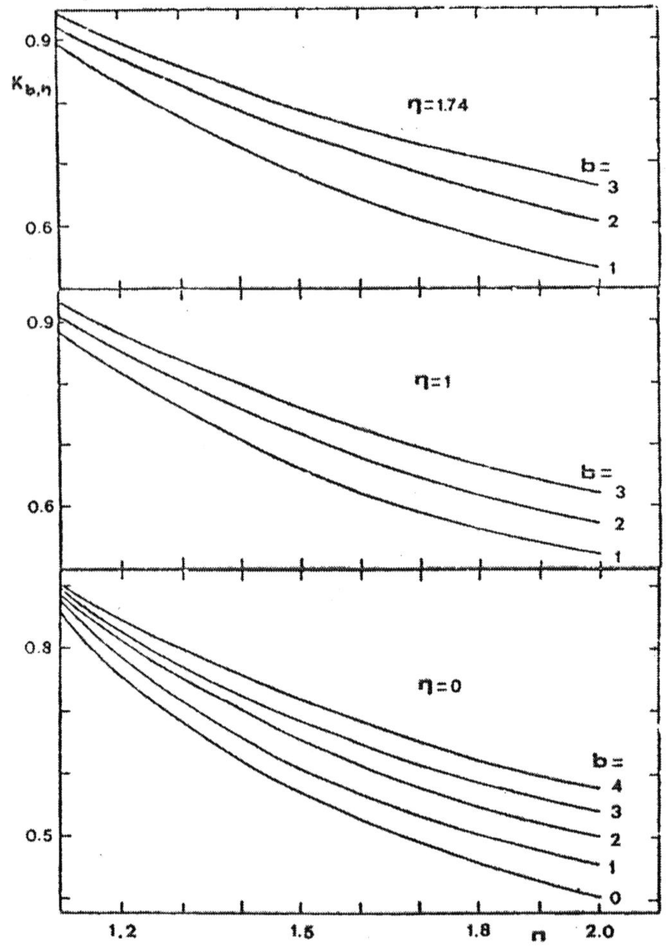

Figure 1

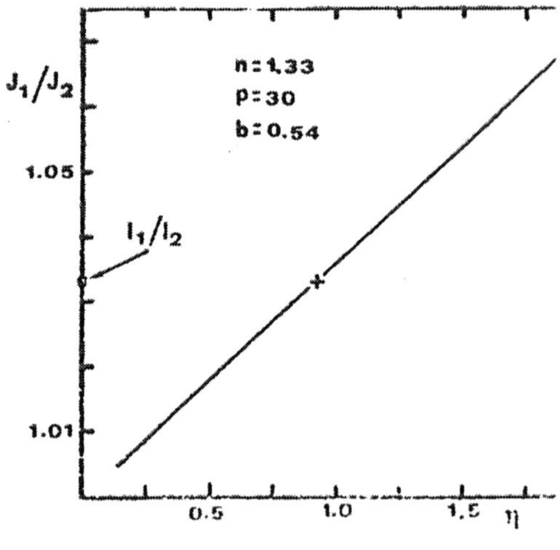

Figure 2

1.D.Karabourniotis and J.J. Damelincourt, J.Appl.Phys. 53, 2965 (1982).
2.D.Karabourniotis, C.Karras, M.Drakakis and J.J. Damelincourt J.Appl. Phys.53,p.72[illegible] (1982)
3.D. Karabourniotis, J.Phys. D, Appl. Phys. (in press).
4.R.D.Cowan and G.H.Dieke, Rev.Mod.Phys. 20 , 418(1948).

n° 40

TEMPERATURE AND LINE BROADENING INFLUENCE ON THE EMISSION OF THE SELF REVERSED LINES IN A HIGH PRESSURE DISCHARGE OPERATED AT A.C. POWER LINE FREQUENCY

Jean-Jacques DAMELINCOURT, Abdelkrim EL HASNAOUI, Dominique GUILHEM and Jacques SALON
Centre de Physique Atomique, Laboratoire Associé au CNRS n° 277, Université Paul Sabatier, 118 route de Narbonne, 31077 TOULOUSE CEDEX, France
Dimitrios KARABOURNIOTIS
Department of Physics, University of Crete, IRAKLION, Greece

Two important parameters in the light emission are the increase of temperature and lowering or radiation trapping by lines broadening. In this work an attempt is made to link the radiation emission with the temperature and an effective width of the lines. In order to determine the importance of the different broadening mechanisms, we have compared the results of a theoretical model with the lines profiles obtained under various experimental conditions.

Figure 1 gives the contribution to the total width of the different broadening effects for the center of the discharge and for two instants of the period.

The knowledge of the local broadening of the line and resolution of radiation transfer equation leads to the observed line profile but the calculations of the total line intensity I_T is very long by this way.

One can write I_T under the form : $I_T = I_{\lambda max}\ \Delta\lambda_{eff}$ where $I_{\lambda max}$ is primarily a function of the temperature and $\Delta\lambda_{eff}$ has the dimension of an effective line width. The experiment performed here show it is possible, at least under the actual experimental conditions to use as $\Delta\lambda_{eff}$ the quantity a + b defined in figure 2.

This figure gives for two lines and two input powers the results obtained from :

$$I_T = I_{\lambda max} * (a + b)$$

and the values obtained from direct measurements of I_T.

Works are actually in progress in order to give a relation between $\Delta\lambda_{eff}$ and the local line profile width.

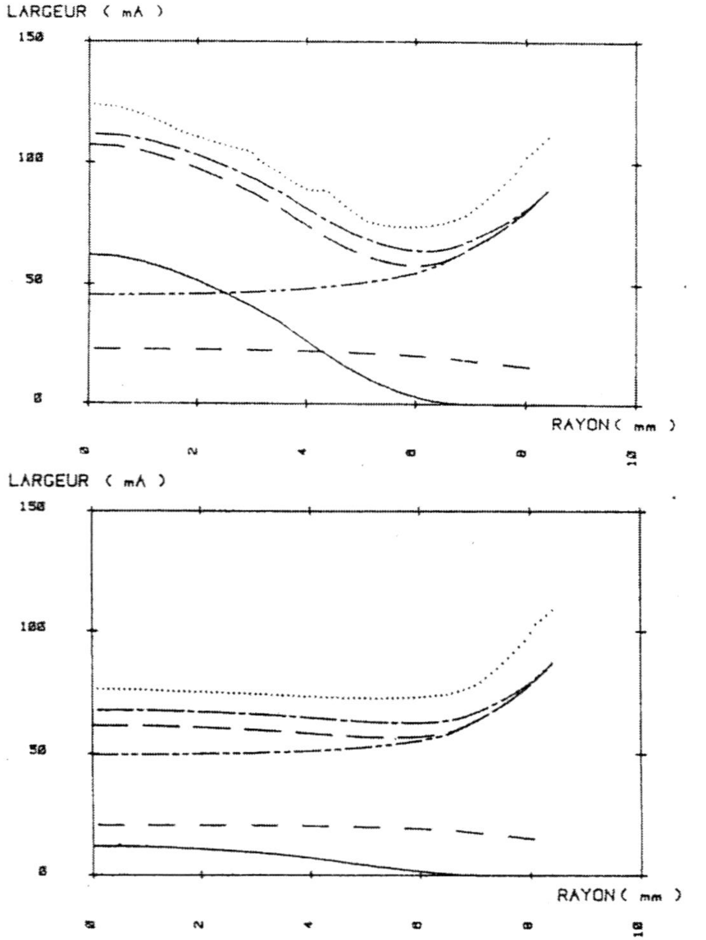

FIGURE 1 - Contribution to the total local width of the different broadening mechanisms (center of the discharge at 0 and 5 ms after the zero of electric field). Mercury discharge. P_{Hg} = 2.7 At, P = 1030 W.

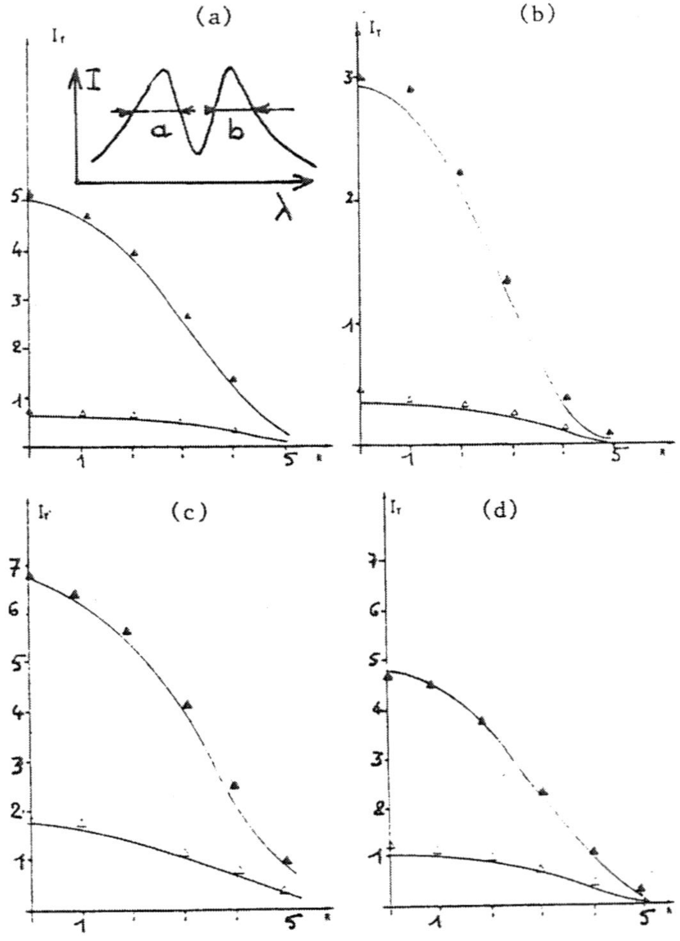

FIGURE 2 - Measured intensities (10^6 erg cm^{-7} s^{-1}) and obtained from $I_{\lambda max}*(a+b)$ at 0 ms (Δ) and 5 ms (▲).
a) 365 nm-1030 W ; b) 365 nm-660 W ;
c) 546.1 nm-1030 W ; d) 546.1 nm-660 W.

n° 41

STIGMATIC COMA FREE DEVICES FOR HIGH RESOLUTION LINE PROFILES MEASUREMENTS IN THE U.V. AND V.U.V. SPECTRAL RANGE.

M. Pouey
Laboratoire de Physique des Gaz et des Plasmas* n Bât. 212, Université de Paris-Sud, 91405 Orsay Cédex, France.

There is indoubtedly a great interest at the present time in U.V. and V.U.V. linewidth and shift measurements for plasma diagnostics and more generally in the studies of atomic and molecular collision process allowing a better knowledge of light sources in term of spectral purity, linewidth and life time. However, as actually most of the available optical devices are characterized by dissymmetrical Line Spread Function, the tail of the image profiles being generated by the coma type aberrations and the astigmatic curvature, it is not easy to use classical deconvolution methods without loss of accuracy.

Actually we consider a device using in a given configuration, a turett with various gratings, each grating covering a given band pass, and allowing high resolution linewidth and shift measurements the device being stigmatic and comas free. Indeed it has been shown (1) that, if we assume that the focusing condition is determined by the conditions for which, <u>independently in both spaces</u>, <u>the third order aberration terms vanish first</u> (instead of the second order one as generally), aberrations terms are identically equal to zero until the fourth order, for a spherical holographic grating with a non uniform grooves distribution.

Assuming that the object is on the grating normal ($n = \sin\delta\ \lambda_o^{-1}$ being the number of lines per mm) as well as C one of the two recording point sources C and D used to generate the hologramm at the λ_o laser wavelength, the grating equation becomes

$$\sin\beta = Pn\lambda_o = P\sin\delta = \frac{m\lambda}{\lambda_o}\sin\delta$$

where β is the angle of diffraction, associate with the working wavelength λ in the m order.

The object could be located, either at the curvature center ($r_c = R$, $\eta = o$) either at $r = R\,(1-P)$, ($r_c = R(P-1)\,P^{-1}$, $\eta = o$), the image locus being a circle of radius $R' = R(2\sin\delta)^{-1}$ passing through the summit and the center of curvature of the grating and centered on the diffracted beam direction determined by $\tan\beta_o = \cot\delta$ (2).

.../...

*Laboratory associated with the C.N.R.S.

The location of the second point source D is then determined in polar coordinates by :

$$\sin\delta = \eta\lambda_o \quad ; \quad r_D = \frac{r'}{P} = R\,(\cos\delta + \sin\delta\ \mathrm{cotg}\ \beta)$$

r' being the image distance.
Moreover, for a given grating, and the polychromatic source as layed down before on the grating normal, the various spectral lines are focused on a plane if $\sin\beta^{x} = 0{,}6077$, r' being approximatively equal to 1.26 R.
For example, at $\lambda^{x} = \lambda_o/3 \simeq 162$ nm ($m = 2$, $n = 1868$ ℓ/mm, $\delta = 69°07$) the limiting resolution associated with the diffraction is about .6 10^{-3} nm for a 100 x 100 mm^2 grating area . If we are using a flat array of silicon CCD imagers overcoated with coronene (3) (pixel size 15 x 15 µm), with a 50 cm radius of curvature grating (r' ∿ 63 cm) a practical resolution of .6 10^{-2} nm should be obtained with an "etendue" of about 6 10^{-3} mm^2 sr.

REFERENCES

(1) M. Pouey, C.R. Acad. Sci., to be published.
(2) M. Pouey, Proc. Int. Conf. SRI 82 (Hamburg) to be published in Nucl. Instr. and Methods.
(3) M.H. Blouke, M.W. Cowens, J.E. Hall, J.A. Westphal and A.B. Christensen, App. Opt., 19 (1980) 3318.

n° 42

CALIBRATION POSSIBILITIES AT THE "LABORATOIRE NATIONAL D'ESSAIS" (L.N.E.) IN THE FIELD OF PHOTOMETRY AND SPECTRORADIOMETRY.

A. JANEST

LABORATOIRE NATIONAL D'ESSAIS - 1, rue Gaston Boissier - 75015 PARIS

1 - INTRODUCTION

The laboratory of "spectroradiophotometry" of L.N.E. is, since 1980, a calibration center of the Bureau National de Métrologie (B.N.M.) The standards are directly calibrated by the Institut National de Métrologie (I.N.M.) which is the primary laboratory in this field.

The diversity of activities by the LNE can displayed as well lamps manufacturers, as laboratories specialised in lightning. They deal with the calibration of light sources, detectors and measuring devices of optical radiations.

2 - PHOTOMETRIC MEASUREMENTS

The following properties can be calibrated : luminous flux, luminous intensity, illuminance and luminance.

2.1. Luminous flux

With a lumenmeter (integrating sphere of 3.5 m, 2 m, 1 m), the luminous flux of a light source is compared to the luminous flux of a standard lamp. The inside of the spheres is coated with a special white paint, spectrally flat. The spectral responsivity of photoelectric receptor of the sphere match very closely the $V(\lambda)$curve. With this apparatus, it is possible to compare sources with very different spectral distribution. Luminous flux between 10 lm and 10 000 lm can be measured with an accuracy of 1 to 5 % according to the level and the source.

2.2. Luminous intensity - Illuminance

With a photoelectric photometer, the illuminance produced by a source is compared to the illuminance produced by a standard lamp. The receptor of the photometer is a high quality $V_{(\lambda)}$ corrected photodiode, with a good linearity and a good $V_{(\lambda)}$ correction. Luminous intensities between 5 Cd and 10 000 Cd can be measured with an accuracy of 1 to 2 % according to the level.

2.3. Luminous luminance

The luminance of a source is compared with a good luminancemeter to the luminance of a lambertian diffusor illuminated by a standard lamp. Luminances between 10 and 2 000 Cd/m^2 can be measured with an accuracy of 3 to 5 %.

3 - SPECTRORADIOMETRIC MEASUREMENTS

Calibrations can be carried out for the sources (spectral distribution) and for the detectors (spectral responsivity).

3.1. Absolute spectral distribution

Spectral distribution measurements can be carried out for spectral irradiance and spectral luminance within a wavelength range of 350 - 900 nm Measurement are made by comparison, wavelength after wavelength, to standard lamps. The measuring apparatus is the same for the two quantities, only the entrance optical system change. A double grating monochromator is used ($\Delta\lambda$ = 0,5 nm). Wavelength exploration and data acquisition are computerized.

The mean accuracy is 5 %.

3.2. Absolute spectral responsivity

Measurements of spectral responsivity are carried out by comparison to standards detectors : silicon photodiodes, chilled germanium photodiode and compensated thermopiles. The wavelength range is 313 - 1 700 nm. For this type of calibration, the L.N.E. use two different apparatus : one is fully automated and measures for 11 differents wavelengths between 400 and 1 150 nm, the other one can achieve calibrations for all wavelengths between 313 and 1 700 nm.

Accuracy is 5 % to 15 % according to the wavelength.

3.3. Color temperature and radiance temperature

Calibration for color temperature can be carried out between 2 042 K and 3 200 K, with an accuracy of 5 K to 10 K. These measurements are made by comparison of the energy ratio of two wavelengths (of a source and a standard lamp).

Calib_ ıtion of radiance temperature can be carried out between 1 100 K and 2 900 K, with an accuracy of 3 to 7 K. Theses calibrations are made by comparison to standard ribbon lamps with a photoelectric luminance comparator working at the wavelength of 650 nm.

o

o o

SPECTRAL PROPERTIES OF METAL HALIDE LAMPS WITH RARE EARTH AND Sc ADDITIVES

P. Tielemans, P. Drop and R. Lorenz*
Nederlandse Philips Bedrijven B.V., Lighting Division
Eindhoven, The Netherlands
* N.V. Philips Industrie,
Turnhout, Belgium

For the application of metal halide lamps in interior lighting low colour temperatures, 2800 to 3800K, and good colour rendering properties, Ra ∿ 80, are required. So far only tin halide lamps fulfilled these requirements. However lamp efficacies of these systems are relative low.

We have investigated the possibilities of Sc, Ce, Pr, Nd, Lu and Tm iodide additives combined with sodium iodide at different molar ratios. This was carried out in a fused quartz discharge tube operated horizontally at 400W. In general we find higher luminous efficacies (η) and lower colour rendering indices (Ra) at higher sodium contents (x_{Na}). For $x_{Na} = 0.5$ η varies from 80 (Sc, Pr) to 100 lm/W (Tm, Ce), for $x_{Na} = 0.9$ η varies from 90 to 110 lm/W. The colour rendering properties are the best for Tm. The higher efficacies at high x_{Na} values are mainly due to a better matching of the sodium radiation to the eye-sensitivity curve. The energy radiated in the visible part of the spectrum remains almost constant with varying x_{Na}, while the colour coordinates lie almost on a straight line. For the elements investigated, these lines intersect the Planckian locus at different points. Ce, the element with the strongest "green aspect", intersects at about 2800K, the other extreme is Tm which intersects at about 4200K.

If the "white lamps" are compared, Ce and Pr show the highest luminous efficacies: 110 - 115 lm/W, Tm and Sc result in 95 lm/W. With respect to the colour rendering properties Tm is superiour: Ra = 80, whereas the others show 65.

An interesting aspect of the spectra of lamps with different Na-contents, is the fact that, within certain limits, the spectral power distribution varies linearly with the mol fraction. This means that the variation of composition in the vapour phase does not lead to strong variations in the discharge conditions. The broadening of the Na-D lines does not influence the linear behaviour. Some experiments were carried out with mixtures of two lanthanides. Again a linear dependence on composition was found. This enables us to predict spectra of lamps filled with mixtures of three or more components, and gives us a new tool for the optimalization of spectra of metal halide lamps. For mixtures of three components this optimalization has been carried out. However we found that the advantages with respect to the binary mixtures are marginal.

As a result a good compromise between η and Ra for interior lighting has been found in a TmI_3, NaI mix. In a 150W version for horizontal operation, where $YVO_4 : Eu^{3+}$ is applied on the outer bulb and a small amount of TlI is added for correction of the chromaticity coordinates, a combination of η = 85 lm/W and Ra = 80 at T_c = 3800K, is realized. For outdoor lighting, where excellent colour rendering properties are of less importance, combinations of CeI_3 or PrI_3 with NaI are more favourable due to the higher luminous efficacies. Preliminary results of life experiments do not indicate that there are fundemental restrictions for application of the various systems in lamps.

n° 44

Investigation of luminous efficacy in Scandium-Sodium iodide arc lamps

M.Saito, S.Taguchi, R.Suzuki, K.Watanabe and M.Tsuchihashi
(Mitsubishi Electric Corporation Consumer Products Research Laboratory,
2-14-40 Ofuna, Kamakura, Japan)

ABSTRACT

Among the various kinds of metal halide arc lamps, Scandium-Sodium iodide arc lamps are occupying an important place as a light source with a high luminous efficacy, because of the high fraction of radiated power in the visible wavelength range. Spectral power distribution emitted from this light source mainly consists of several spectral lines from Hg, D-lines from Na, and a large number of lines radiated from Sc. The radiation from the additives is significantly affected by chemical and physical processes, which are roughly classified into two steps. One is thermochemical properties of ScI_3-NaI molten salt which determine the vapor pressure and composition over the molten salt. The other is axial and radial energy flow in the arc column. Concerning the thermochemical properties, solution ideality of ScI_3 - NaI molten salt and formation of both complex compounds and polymers must be considered. Furthermore, the molten salt is known to creep along the surface of the arc tube, so that temperature distribution of the molten salt must be also considered for precise study of vapor composition because activity is dependent upon the temperature. On the other hand, concerning the energy flow, such an theoretical arc model is needed including transport of reaction energy due to diffusion fluxes, heat conduction fluxes, radiation fluxes, axial convection, and axil and radial demixing effects. A number of improvements are still required to establish the accurate arc model.

This paper describes the relationships between several factors for lamp design and radiation characteristics of the lamp. As a first step, we have investigated the dependence of radiation power from Hg, Sc and Na on the above factors, such as mole fraction of $NaScI_4$ in $NaScI_4$ -NaI molten salt , amounts of ScI_3 -NaI additives, relative cold spot temperature of condensed molten salt, diameter of arc tube and wall loading. While the above mole fraction, amounts of additives and the cold spot temperature affect the vaporization properties of the molten salt, diameter of arc tube and wall loading primarily affect the axial and radial energy flow. The latter factors are also accompanied by the influence on the cold spot temperature.

The overall luminous efficacy can be split into the visible radiant efficiency and the luminous efficiency of the visible radiation. It is interesting fact that the overall luminous efficacy is roughly dependent upon the visible radiant efficiency, and the visible radiant efficiency is proportional to the sum of radiation power from Sc and Na over our experimental range. Furthermore, radiation power from Na is predominantly affected by changing the above factors, compared with radiation power from Sc. This tendency implies certain effects of demixing of Na particles.

As shown in past literatures, increase in wall loading leads to significant increase in luminous efficacy because of realizing higher cold spot temperature and higher input power in unit arc length. In such a sense, increase in wall loading might be more effective in achieving higher luminous efficacy, but this would result in a serious decrease of lumain maintenance of the lamps. We will discuss the possible method to improve the luminous efficacy in Scandium-Sodium iodide arc lamps without increasing wall loading.

DEVELOPMENT OF A SEALED BEAM PHOTOPRINTING LAMP CONTAINING IODIDES OF IRON AND GALLIUM

AUTHOR - R. B. Page.

THORN EMI Lighting Limited, Melton Road, Leicester, England.

CONCEPT

The availability of the PAR 64 sealed beam outer lamp jacket, originally developed for aircraft landing systems, has stimulated a similar concept for a photoprinting source. If a reflecting dichroic layer is applied to the inner surface of a PAR 64 outer jacket, and a discharge tube is positioned close to the primary focus of the reflector; then by diffusing the radiation through a prismatic front window, a uniform directionally controlled beam is produced which will satisfactorily irradiate large photographic plates. Additional benefits of this arrangement include the efficient dispersal of heat through the dichroic layer and the high degree of protection from short wave UV radiation provided by the glass outer envelope.

The sealed beam design encourages the development of a short, highly loaded, arc tube. As lamp life for a photoprinting source is primarily dependent on the number of switching cycles, high wall loadings of up to 37 W/cm^2 have been used achieving up to 10,000 exposures per lamp. An additional feature of the lamp is the hot restart potential which, if combined with a fast run-up characteristic, avoids the need to simmer at reduced power, thus achieving greatly reduced running cost when compared with a conventional linear source.

The discharge employs iodides of both iron and gallium which, together with mercury, produce the desired radiation. Both iron and gallium systems have been examined separately, and in combination, and the complementary nature of their respective spectrums is illustrated in Fig. 1. The iron spectrum is particularly useful for exposing photosensitive polymer emulsions which favour wavelengths of 360 - 400 nm. Iron also serves to substantially boost the 400 - 450 nm region, thereby enhancing the gallium response which is traditionally used for exposing diazo photosensitive materials. Fig. 1. also emphazies the sharp UV cut-off characteristics of the glass outer jacket.

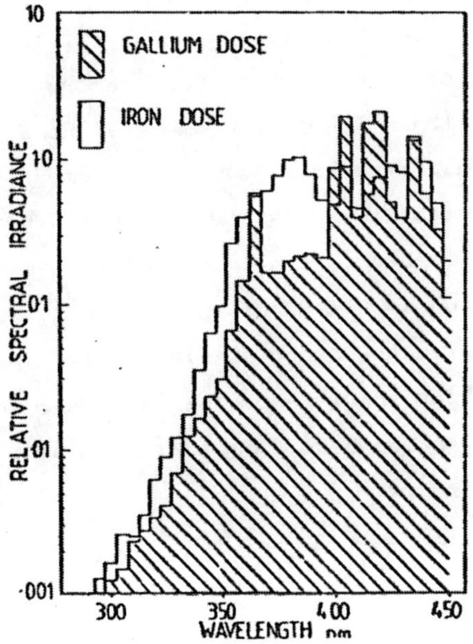

Fig.1

EXPERIMENTS

A GaI_3-Hg system was examined in which the Hg content is held constant whilst the GaI_3 quantity was varied. The UV.A and effective photoprint response (300 - 450 nm) was measured using a 'Macam' UV radiometer on a screen positioned 1 metre from the source.

Similarly, a FeI_2-Hg system was also examined. In this case it should be noted that the Hg content was progressively reduced as the FeI_2 dose weight was increased. This was found to be necessary because, increasing the FeI_2 quantity with a constant Hg dose weight resulted in an excessive increase in lamp volt drop, thereby making operation from a 240V mains supply, very unstable.

Both systems indicated a peak response at a determined iodide dose. This was particularly noticable with iron and the knowledge was used to investigate a combined FeI_2-GaI_3- Hg system. In this experiment the FeI_2 and Hg quantities were held constant whilst the GaI_3 additive was progressively increased. Again, a pronounced optimum response was observed.

PERFORMANCE

Finally, a comparison was attempted between an optimised 800W sealed beam lamp and a conventional 2 kW, GaI_3 linear source, operating in a typical photoprint housing. Irradiation measurements were taken at 1 metre in direct line with the beam centre using a 'Bentham' spectroradiometer. Results are illustrated in Fig. 2. The comparison suggests substantial gains for the sealed beam lamp at lower power, in all the relevant spectral regions. Distribution of screen radiation (centre to corner readings) give values of over 95% for the linear embodiment and above 80% for the sealed beam lamp operating in a suitable luminaire.

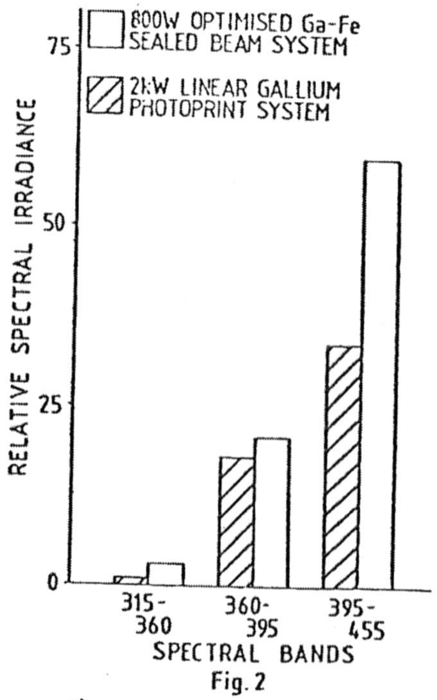

Fig. 2

n° 46

HID LAMPS INCORPORATING A HEAT PIPE

F L Whittaker
Research and Engineering Laboratories, THORN EMI Lighting Limited, Melton Road, Leicester, England.

The lamp industry is continually striving for ways of increasing the effective vapour pressure of refractory dose constituents in HID lamps. In the present work, the effective vapour pressure above various liquids is increased through the use of capillary action to transport the liquid against a temperature gradient. Evaporating liquid can be returned to the liquid pool after condensation and a cycle set up as in a heat pipe. Three requirements must be met:

(1) a porous wick must be provided along a temperature gradient

(2) the dose material must form a liquid capable of wetting the wick

(3) the dose liquid must make contact with the wick.

In the simple examples described, requirement (1) was met by replacing the overwind of one electrode by a tungsten wick or sponge and allowing this to extend to the end of the arc tube. The wick can take many forms: a braid has been found very suitable. Requirement (3) was met by operating the tube vertically with the wick at the bottom and shaping the arc tube end so that the liquid dose would form a pool around the wick. Gravitational forces were thus used to make the contact and the capillary forces had to oppose gravity as well as the temperature gradient. Requirement (2) has to be tested by trial. With metal halide lamps it can be assumed that the mercury evaporates completely, so that wettingdepends only on the metal halides. Silica arc tubes with a wick at one end were dosed with excess sodium iodide. When operated with the wick at the top, the appearance was typical of a sodium iodide lamp, with the sodium emission showing itself as a modest orange halo around the arc core. On inverting the lamps so that the wick was at the bottom, the orange colour became much deeper, the luminous efficacy increased and it was possible to obtain instabilities of the kind observed in metal halide lamps when the halide pressure is excessive. This behaviour was reproduced in subsequent runs in each orientation. The method can be extended to other halides which liquefy at typical arc tube temperatures but have low vapour pressures.

Another interesting case is where the dose is a metal or metal alloy, especially in arc tubes of polycrystalline alumina. Mercury, a normal constituent of such lamps, is notoriously reluctant to 'wet' wicks: its contact angle is 155° with alumina and 130° with tungsten. However, it appears that the wetting of tungsten depends on the reduction of an invisible layer of WO_3 on its surface, and this can be achieved by (for example) liquid sodium at temperatures exceeding $160^{o}C$. High pressure sodium lamps were made with a wick at one end and specially shaped ends to promote contact between the amalgam and the wick. They were dosed with sodium amalgam (20% Na w/w) and operated vertically with a heat shield at the top of the arc tube. The amalgam was found to wet the wick satisfactorily with a large increase in sodium partial pressure when the heat pipe was in use. The characteristics of otherwise identical lamps are shown in the Table:

	Arc Voltage	Current(A)	Arc Watts	CCT	R_a
Without heat pipe	106	4.3	400	2000K	10
With heat pipe	385	1.3	400	3100K	65

The extent of vapour pressure enhancement can be controlled by changing the length of wick. The large increase in effective cool spot temperature achieved with the heat pipe makes possible the manufacture of metal vapour lamps containing metals other than sodium or mercury.

n° 47

DEVELOPMENT OF A LONG LIFE INCANDESCENT LIGHT BULB FOR SPACE APPLICATIONS.

R. L. Roberts and N. Helmold

Santa Barbara Research Center, Goleta, California, U.S.A.

and National Aeronautics and Space Administration, Goddard Space Flight Center, Greenbelt, Maryland, U.S.A.

This paper discusses the selection of filament material and operating conditions for an incandescent light bulb for use in a long life (in excess of 40,000 hours) high reliability space environment.

An optical encoder with a 1.75 watt incandescent light bulb is used to provide information for angular position control of an object-space scan mirror on meteorological satellites. The light bulb, operated at a constant five volts (V) direct current (DC) in a krypton atmosphere, is coiled in a single helix. The light bulbs for the first encoders met the 40,000 hour mission requirement but the manufacturer, at some time, made unrecorded changes that shortened the life to approximately 8,000 hours. At the onset of these premature failures, a joint investigation was begun by the Santa Barbara Research Center (SBRC) and the National Aeronautics and Space Administration (NASA) Goddard Space Flight Center (GSFC). The investigation focused on the following:

1) Determining the cause of failure,
2) determining the optimum operating conditions and material for the filament,
3) determining screening criteria.

Item 1) - Three separate failures were found. The first was oxygen contamination as evidenced by a "christmas tree" growth of tungsten oxide on the filament. This failure was traced to a rigid potting material at the base of the envelope which caused the envelope to crack under the stress of temperature. A second failure caused by "water cycling" was noted by the darkening of the envelope after a few hours of operation.

This problem was eliminated by vacuum baking the envelope prior to backfilling with krypton. The third failure occurred after approximately 8,000 hours of operation. This failure was one in which the filament separated near its center. Analyses by Scanning Electron Microscope (SEM) of the surface and cross section revealed a regular pattern of facets and a single crystal spanning many turns, which therefore led to the conclusion that the wire was probably "pure" tungsten. The lack of specific knowledge of the wire type that was used in the past to provide filaments lasting in excess of 40,000 hours led to a program to regain the required life.

Item 2) - To determine the optimum filament type and operating conditions, light bulbs were made with filaments using tungsten and tungsten three percent rhenium from Luma Corporation of Sweden, and tungsten three percent rhenium and tungsten one percent thoria from General Electric. Five bulbs from each group are being tested at 5 V DC, 3.6 V DC, 5 V alternating current (AC) and 3.6 V AC. Current and light output decay are being monitored and compared with a reference sample of five bulbs made from the same filament material which failed after 8,000 hours. In addition, SEM's of the filament surface and longitudinal cross sections are being analyzed after 500 and 1,500 hours of operation. This data, with relative temperature profiles, will be used to select a filament material and operating current for use in future encoders.

Item 3) - An effort was begun by NASA/GSFC to predict the life time of of the light bulbs presently in orbit. This study consists of several groups of bulbs made from the same lot as those presently in orbit (expected to last approximately 8,000 hours at 5 V DC). These bulbs are being run at various DC voltages so that the exponent that describes the light output versus life time can be calculated. Presently there are three instruments in orbit, two operating at a fixed 5 V DC and the other operating with a voltage that can be incremented from 4.45 to 5 V DC.

n° 48

Power saving incandescent lamp with infrared-reflecting filter

H. Köstlin, B. de Vrijer*)
Philips GmbH Forschungslaboratorium Aachen, Germany
*) Philips Centraale Ontwikkeling Licht, Eindhoven, Netherlands

The design of a lamp has been treated theoretically and experimentally with the aim to increase the luminous efficacy by means of a filter on the inner side of the bulb, which is transparent in the visible and reflecting in the infrared. The intensity and local distribution of the infrared radiation focussed back onto the filament were studied by a finite element analysis of the filament radiation with respect to the shape of bulb and filament and to the influence of misalignments. A mains voltage lamp of common geometrical dimensions with its rather long and thin filament is found to give highest power savings, if the reflector is elliptic and the filament linear and positioned in between the focal-points. The zones of the reflector which are close to the axis hardly contribute to the power saving and thus the openings were introduced here for the filament supports and for a dome.

The equations were derived which allow the calculation of the gain in luminous efficacy as a function of the reflectance and transmittance of the filter and of the emittance of the filament, taking multiple reflections into account, and were used to optimize the filter. The filter is based on a thin Ag film which is sandwiched between dielectric films. High infrared reflectance requires a Ag film as thick as possible. High visual transmittance (T_V), however, can only be induced into the metal film by the dielectric films up to a certain Ag thickness. The refractive index of the dielectric films must be as high as possible, i.e. of the order of the visual absorption index of Ag (k = 3.4). The optimal thickness of the dielectric films acting at $\lambda = \lambda_V$ as anti-reflection coatings of the Ag film is at about d $\gtrsim$ $\lambda_V/(8n)$. The visual reflectance of the filter (R_V) is not entirely required to be minimized, as light primarily reflected in the lamp housing also can contribute to the light output of the incandescent lamp. Hence the effictive transmittance of the system is, besides other factors, proportional to $T_V/(1-R_V)$, a quantity which is sometimes called net-flux-ratio of the filter and shows that only absorption is essential for visual losses. Most critical is absorption in the Ag film, which is not only due to the normal bulk absorption of Ag but also due to imperfections of the very thin ($\approx$ 0.02 µm) Ag film.

Lamps were made from elliptic bulbs of dimensions similar to 60 W GLS lamps ($\approx$ 60 mm ø). They were inside coated with a ZnS/Ag/ZnS filter by evaporation and filled with Kr/N_2. For the same filament temperature and a life time of 2000 h an increase in luminous efficacy of 30 % and more was found compared to lamps without infrared-reflecting filter.

n° 49

Energy Saving Tungsten Halogen Lamp with Infra-red Reflective Films

Noboru ICHINOSE, Kiyokazu HONDA, Yooji YUGE,
Ariyoshi ISHIZAKI, Teruo OHSHIMA
Lamp Division, Toshiba Corporation
1-201-1, Funakoshi-cho, Yokosuka-shi,
Kanagawa 237 JAPAN

Conventional incandescent lamps including tungsten halogen lamps emit only 10% visible light while burning. The rest of energy are consumed to be emission of infra-red rays and heat loss through lead-in wires. The idea utilizing infra-red rays effectively was introduced by MIT (Massachusetts Institute of Technology, USA) Group in 1974[1], and applied for incandescent lamps[2]. They applied thin films composed of TiO_2 - Ag - TiO_2 to an inner surface of a bulb which reflected infra-red rays back to a filament. However, these TiO_2 - Ag - TiO_2 films were lack of heat resistance at the temperature of more than 200°C.

Meanwhile, we have been developing infra-red reflective films which held heat resistance.

Here, we report these films with high heat resistance up to 1000°C based on fine chemical technology and their application to tubular tungsten halogen lamps.

These films consist of multi-layers TiO_2 - SiO_2 films which reflect infra-red rays effectively.

When we compose 9 layers TiO_2 - SiO_2 ... TiO_2 on an outer surface of a bulb of the tungsten halogen lamp J100V300W, the performance of the lamp is greatly improved in its efficacy and ratio of heat ray reduction. They are tabulated in Table 1 in comparison with the conventional one.

The efficacy is improved by 23.9%, and the heat rays are reduced by 46%.

J 100 V 300 W		
	with Infra-red Reflective Films (TiO_2-SiO_2 9 layers)	No Film
Efficacy (Lm/W)	22.3	18
Ratio of Heat Rays Reduction	0.54	1

Table 1

Applying this technology to tublar tungsten halogen lamps, we expect that the wattage of those lamps can be reduced by more than 20%, and heat rays can be reduced by more than 60% at the same time.

Reference:

1) J. C. C. Fan et al.
Proc. of 9th Intersociety Energy Conversion Engineering Conf. San Francisco 341 (1974)

2) J. Brett et al.
IES paper No.15 (1979)

n° 50

Theoretical Description of Tungsten Transport in Halogen Incandescent Lamps

E. Schnedler
Philips GmbH Forschungslaboratorium Aachen, Germany

A numerical finite difference procedure is presented which allows the calculation of transport currents in linear, cylindrical halogen incandescent lamps, taking into account the different bulb temperatures at different sites. Assuming local chemical equilibrium the model fulfils the mass balance equations including concentration diffusion and thermal diffusion terms at each node point. Reactions with condensed phases at the filament and at the bulb are allowed.

The first example is concerned with an incandescent lamp filled with 10^{-2} bar Br and 5 bar Kr only. It turns out that this lamp will blacken after a short time due to the condensation of solid lower tungsten bromides on the bulb. Thus the bromine contents of the gas phase is reduced until deposition of tungsten at the bulb wall appears.

This theoretical result is in agreement with the experimental verification of this type of lamp which shows blackening after a few hours of lifetime.

A second example deals with a halogen lamp filled with the commonly used CH_2Br_2-Kr filling. It turns out that the carbon contents of the gas phase will be reduced by the formation of solid tungsten carbides at the cold leads and thus is lost as a gaseous getter for oxygen. Therefore impurities of oxygen in the lamp volume have a major influence upon the magnitude of the tungsten transport currents from hot parts of the filament to cold ones by the formation of gaseous oxygen-containing tungsten compounds. Predicted transport behaviour and lifetimes are in agreement with experiments.

As a third example a system with a transport direction from cold to hot sites of the filament, the W-O-Cl-system, will be discussed. The gas filling of 10^{-2} bar Cl and 10^{-3} bar O is reduced by the formation of solid tungsten oxides and tungsten chlorides. The theory predicts large tungsten transport currents from the cold ends of the leads to the hot filament. From the magnitude of the transport currents a lifetime of a few hundreds of hours can be calculated which again is in agreement with the experimental verification of this type of incandescent lamp.

Fig. 1 shows the tungsten transport current density for a cylindrical incandescent lamp filled with 5 bar Kr, 10^{-2} bar CH_2Br_2, and an impurity of 10^{-4} bar O.

In contrast Fig. 2 shows the tungsten transport currents in a lamp filled with 10^{-2} bar Cl and 10^{-3} bar O.

(Transport currents are measured at the small crosses. Direction and magnitude are given by the direction and length of the lines).

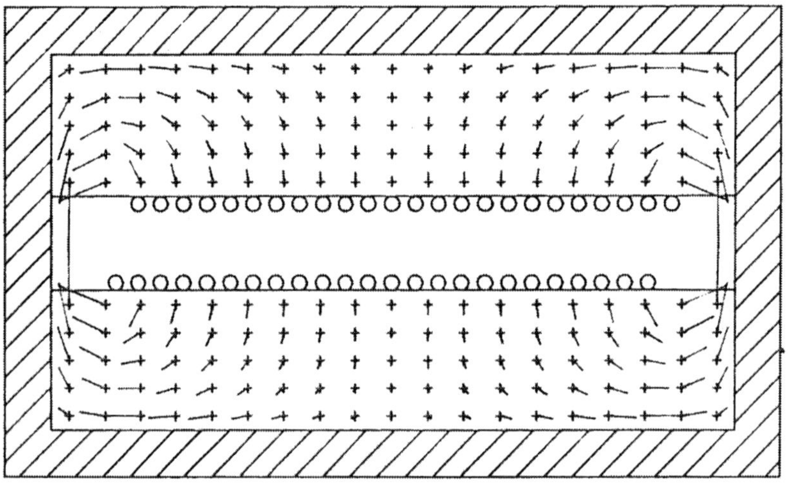

Fig. 1 W-current-density: ⟶ 2.68 E13 particles/(cm*s)

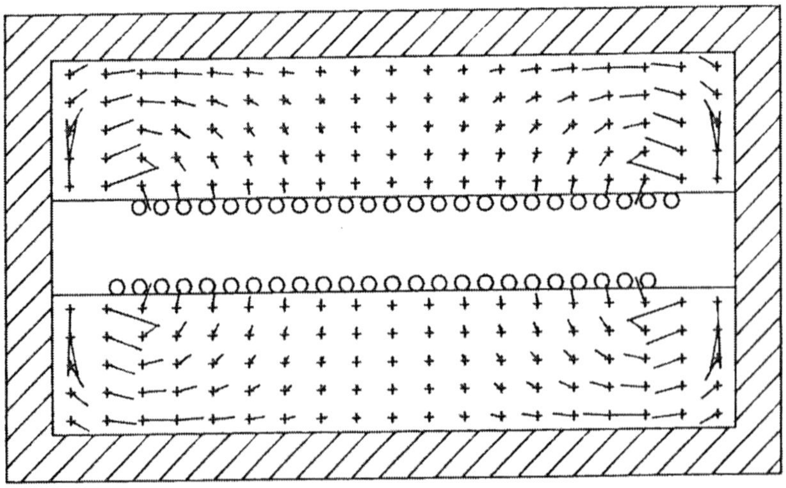

Fig. 2 W-current-density: ⟶ 4.58 E14 particles/(cm*s)

PREDICTIVE MODEL FOR THE LIFE OF GAS FILLED INCANDESCENT LAMPS

Rolf S. Bergman
General Electric Company
Lamp Business Group
Nela Park
Cleveland, Ohio

Published work on the mechanism responsible for the finite life of incandescent tungsten filaments generally accept the hot-spot model as the best explanation. This model assumes that a local defect along the filament causes an enhanced temperature zone which accelerates evaporation leading finally to lamp failure through localized melting. Summaries of hot-spot model results show rather good correlation between experiment and model predictions.(1) However, we are unaware of any published work which uses the hot-spot model to obtain predictive information useful in incandescent lamp design.

It is the object of the present study to establish a predictive model for the life of coiled tungsten incandescent filaments in a gas environment. To this end, it is assumed that the temperature of the filament is such that tungsten transport properties are dominated by evaporation rather than surface diffusion. For this temperature range, it is further assumed that evaporated tungsten is limited by diffusion through a given gas sheath, assumed to be cylindrical. Calculation of the tungsten transport rate accounts for the dependence of filament life on filament and bulb wall temperature, the gas type and pressure, and coiling parameters. The last factor is important since diffusion from the surface of the coiled-coil takes place from the wire surface. The mass transport rate derived by Covington and Ingold(2) is used. The thickness of the gas (Langmuir) sheath needed in the transport equation is obtained from results of free convection studies on horizontal (vertical) cylinders.

The calculation of the tungsten (and heat) transport properties accounts for all the variables assumed important in determining filament life except for the dependence on wire diameter. The filament diameter dependence is probably the second most important parameter (next to the temperature). Not only does it have a strong direct effect but also includes possible information on defects which cause the development of the hot spot. Information on the effect of wire diameter on life was obtained empirically. A series of 120 volt A-line lamps (40 to 1000 watt) filled with 600 torr argon plus 5% nitrogen were photometered and burned to end of life. From the photometry data the average true temperature of each filament was obtained. From the known life and filament temperature, the filament life at 2800K was calculated using the known temperature dependence of the tungsten evaporation rate. The resultant normalized life at 2800K was plotted against wire diameter. Curve fitting the data shows that filament life can be expressed by

$$\text{LIFE} \propto \exp(-160/d)$$

for d less than about 120 μm, where d is the wire diameter in μm. This surprising result indicates that for small wires, defects of a fixed size are responsible for hot-spot formation.

A computer program was written to simulate the development of an initial hot spot of a given length with time. The results of the computer simulation leads to an equation which adequately predicts the life of incandescent lamps over a large range of parameters. Information on the size of the defects or temperature differences needed to match the results of experiments are also obtained.

REFERENCES

1. E. Fischer, J. Fitzgerald, W. Lechner and W. Lems, Philips Tech. Rev., 35, 296 (1975).
2. E.J. Covington, J.H. Ingold, JIES, 4, 198 (1975).

n° 52

Growth of potassium-filled bubbles in doped tungsten and its relation to hot spot development and intergranular fracture

S. Garbe and S. Hanloh
Philips GmbH Forschungslaboratorium Aachen, Germany

Migration and growth of potassium-filled bubbles under the influence of temperature gradients and stress fields are considered as one failure mechanism of incandescent lamp coils. A model is presented which assumes initial growth of bubbles due to the size dependent bubble migration in the temperature and stress field by volume diffusion of tungsten vacancies.

Bubble size distributions have been analyzed in single coil doped tungsten as function of temperature and time. Also the temperature profile of the coils was measured during life, in order to detect bubble induced hot spot development and grain boundary sliding and fracture.

The size of the bubbles grows in time by volume diffusion as a function $t^{1/2}$. The largest voids develop by diffusion in the radial temperature gradient of thick wires. The bubble size distribution is determined by the mean potassium content of the wire and its recrystallization structure, influenced by solution and grain boundary precipitation of other impurities (segregated third phases caused by e.g. oxygen or selenium in the lamp).

The local temperature rise of the part of the tungsten coil, where bubbles of a certain size and distribution had grown, was calculated by numerical solution of the power balance equation.

Growth of detrimental bubbles was observed by monitoring the temperature profile during life and determination of the cross section profile at the end of life. Bubbles which caused a change of the position of the hottest spot prior to end of life or even intergranular fracture were analyzed.

The mechanism of bubble induced hot spot growth is only life determining for wires of large diameter, large temperature gradients, and long lifetime.

Precipitated potassium in a bubble was observed by X-ray fluorescence after intergranular fracture at room temperature. This proves that the potassium filled bubble growth mechanism is responsible for intergranular fracture at very long lifetimes.

n° 53

THE RELIABILITY CHARACTERISTICS OF TUNGSTEN HALOGEN LAMPS SUBJECTED TO ACCELERATED LIFE TESTS

* D. M. Barry; Lakehead University, Thunder Bay, Ontario, Canada.
* D. K. Das; Marconi Company, Chelmsford, U. K.

SUMMARY

Miniature, single-ended Tungsten Halogen lamps with a rated life of 2000 h at a nominal output of 900 lumens, have been subjected to accelerated life test, the accelerating stresses being overvoltage and intermittent switching.

The life of these incandescent filament lamps is inversely proportional to the evaporation rate of the filament, which, itself, is directly proportional to the absolute temperature achieved by the filament. Filament temperature is a direct function of operating voltage, hence lamp life is inversely proportional to operating voltage. The specific characteristic property of halogen lamps being long life, it was decided to investigate their reliability by the use of overvoltage and switching tests. A power-rule method of relating life at accelerated voltage to life at a designated voltage has been developed.

The lamps were tested, i.e. operated to failure, under the following conditions: (a) operating voltage level at steady d.c. at 50%, 40%, 25%, 10% above rated and at rated voltage; (b) operating voltage intermittent d.c. at the same overvoltage levels as (a) above. For the intermittent tests the cycle time σ = Ton/T was 79%.

The failure data was analyzed using the Weibull distribution, a statistical distribution of wide practical applicability which generates a shape and a scale parameter. The scale parameter, α, refers to the characteristic life of the lamps and the shape parameter, β, is a measure of lamp wear out. These parameters were estimated using Maximum Likelihood Estimation.

A comparison of the failure patterns of the lamps submitted to intermittent tests with those submitted to the steady d.c. tests showed that operating life decreased

* Formerly of the Postgraduate School of Industrial Technology, University of Bradford, U. K.

with increased switching but this decrease was not large in magnitude, 4% being a typical value. The shape and scale parameter at the rated voltage and ideal environmental conditions were estimated at α = 11,758.74 and β = 3.59. It was found that the scale parameter continuously changed with change in operating stress level, whereas the shape parameter remained relatively constant.

A technique known as "equivalent scale parameter" has been developed in order to estimate reliability under hazard conditions from the reliability information estimated during this study. A reliability chart relating burning hours and reliability level for different values of scale parameter has been generated. From this, and the fact that an almost constant relationship was found to exist between the mean and the Weibull scale parameter, it is possible to estimate Weibull reliability against time for raw failure data.

It is anticipated that this technique would be particularly useful to lamp manufacturersand users in formulating a reliability model in a situation where the environmental stress conditions are non-deterministic in nature.

n° 54

NEW Tb^{3+} AND Mn^{2+} ACTIVATED PHOSPHORS AND THEIR APPLICATION IN "DELUXE" LAMPS

J.Th.W. de Hair and J.T.C. van Kemenade
Nederlandse Philips Bedrijven B.V., Lighting Division
Eindhoven, The Netherlands

New green (Tb^{3+}) and red (Mn^{2+}) emitting phosphors with high efficiency when excited with UV have been developped (1). These phosphors show promising application possibilities in three band (Tb^{3+} phosphor) and "Deluxe" (Tb^{3+} and Mn^{2+} phosphor) fluorescent lamps (both with increased efficacy as compared to the existing lamps).

The phosphors are based on a borate host lattice with the general formula $LnMeB_5O_{10}$ (Ln=Y,La,Ce,Gd,Tb; Me=Mg,Mn,Zn). The Tb^{3+}-activated phosphors show the green luminescence characteristic of Tb^{3+}, the Mn^{2+}-activated compounds show broad band red luminescence with a maximum at 630 nm (see fig. 1). By coactivation with Tb^{3+} and Mn^{2+} both emissions are obtained in one phosphor.

Both activators are sensitized by Ce^{3+} for 254 nm excitation. Fig. 2 illustrates the dependence of the quantum efficiency (q.e.) on the activator concentration for phosphors with different $LnMgB_5O_{10}$ host lattices. In $GdMgB_5O_{10}$ relatively low activator concentrations are sufficient to achieve high q.e.'s. This is related to the intermediate role of Gd^{3+} in the energy transfer from Ce^{3+} to Tb^{3+} and Mn^{2+}, as described before (2,3).

The phosphors show a good maintenance behaviour in fluorescent lamps(36W T8): the lumen output at 1000 hr amounts to 98% of that at 100hr. Table I gives additional characteristics of single-phosphor lamps. The luminous efficacy of $(Ce,Gd,Tb)MgB_5O_{10}$ is higher than that of $(Ce,Tb)MgAl_{11}O_{19}$, the green component in three band fluorescent lamps. The new phosphor (CBT) will improve the lumen output of these lamps.

The Mn^{2+}-phosphor is particularly suited for application as red component in "Deluxe" lamps because of the favourable spectral energy distribution of its luminescence. Besides this phosphor enables the step towards more efficient small diameter "Deluxe" lamps since its maintenance is superior to that of the phosphor used at present (tin-activated strontium magnesium orthophosphate).

Table II gives characteristics of some "Deluxe" lamps (36W T8) which are feasible with the new phosphors. Using a Tb^{3+} and Mn^{2+} coactivated phosphor (CBTM) with proper emission colour, a lamp with a colour temperature T_c of 2800K is realized.

Lamps with T_c > 2800 K are realized by application of a blend of a CBTM phosphor and a conventional Ca-halophosphate. Examples of lamps with T_c of 2900 and 3800 K are given. The CBTM-containing lamps have colour rendering properties (R_a and R_9 > 80) comparable to those of the existing "Deluxe" lamps. The luminous efficacy of the new lamps (36W T8), however, exceeds that of the latter (40W T12) by about 30%.

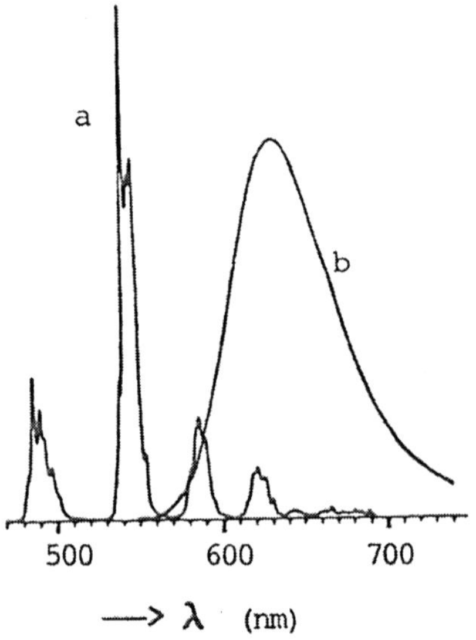

Fig. 1

Emission spectra of

a. $Ce_{0.05}Gd_{0.85}Tb_{0.1}MgB_5O_{10}$ and
b. $Ce_{0.05}Gd_{0.95}Mg_{0.98}Mn_{0.02}B_5O_{10}$ at 254 nm excitation

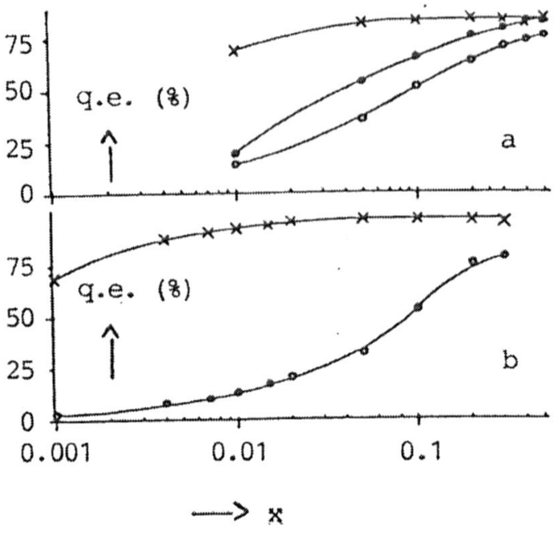

Fig. 2

Quantum efficiencies (q.e.) at 254 nm excitation as a function of the activator concentration x for:

a. Tb^{3+} in $Ce_{0.05}Gd_{0.95-x}Tb_xMgB_5O_{10}$ (x-x), $Ce_{0.05}La_{0.95-x}Tb_xMgB_5O_{10}$ (o-o) and $Ce_{1-x}Tb_xMgB_5O_{10}$ (●-●).
b. Mn^{2+} in $Ce_{0.05}Gd_{0.95}Mg_{1-x}Mn_xB_5O_{10}$ (x-x) and $Ce_{0.05}La_{0.95}Mg_{1-x}Mn_xB_5O_{10}$ (o-o).

Table I Characteristics of single-phosphor lamps (36W T8)

Phosphor	efficacy (lm/W) 100 hr	chromaticity coordinates x/y
$(Ce,Gd,Tb)MgB_5O_{10}$	129	0,324/0,535
$(Ce,Tb)MgAl_{11}O_{19}$	126	0,314/0,542
$(Ce,Gd)(Mg,Mn)B_5O_{10}$	40	0,546/0,301

Table II Characteristics of new "Deluxe" lamps (36W T8)

Phosphor	T_c(K)	efficacy (lm/W) 100 hr	R_a	R_9
CBTM=$(Ce,Gd,Tb)(Mg,Mn)B_5O_{10}$	2800	67	80	82
CBTM + C_1 *)	2900	67	83	95
CBTM + C_2 *)	3800	69	87	86

*) C_1 and C_2: Ca-halophosphates, used in Cool-White and Daylight standard lamps.

References

1. B. Looye, J.Th.W. de Hair and C. Bakker: USP - 4319161.

2. J. Th.W. de Hair, J. Luminescence 18/19 (1979) 797.

3. J.Th.W. de Hair and W.L. Konijnendijk, J. Electrochem. Soc. 127 (1980) 161.

n° 55

RARE EARTH ACTIVATED NIOBATES

W. A. McAllister
Westinghouse Electric Corp.
One Westinghouse Plaza
Bloomfield, New Jersey 07003

Luminescence of calcium niobate, cadmium niobate and their solid solutions has been examined, the intensity of the blue (470nm peak) band generally increasing with cadmium concentration on excitation in a band peaked at 285nm. Emission intensity of several of these phosphors exceeds that of a commercial, Blue Halo powder having a similar spectral distribution under 254nm excitation.

Upon activation by some rare earths (Ho,Er,Sm,Tb) with sodium for charge compensation, both niobate and rare earth line emissions appear, excitation being through the 285nm excitation band of the niobate group and by overlap of niobate emission with the region where f-f transitions occur. Better rare earth intensities arise on inclusion of dysprosium or europium in the matrix, the former being largely responsive to the niobate excitation band, the latter through the niobate emission. For either of these phosphors the niobate emission is almost completely quenched indicating efficient transfer to the lanthanides. Under 254nm excitation the dysprosium activated phosphor has the usual yellow and blue lines but of greater intensity than that of calcium yellow phosphor at the emission maxima; with europium activation however, the 612nm emission intensity is only a third that of a Y_2O_3:Eu standard. These marked differences in luminescence of both host and rare earth activated materials is associated with the influence of cadmium on the structure and thus, energy transfer processes in the phosphor.

CONSIDERATIONS OF COILING IN THE DESIGN OF FLUORESCENT LAMP ELECTRODES

E J Covington (General Electric Co., Fluorescent Systems Department, Nela Park, East Cleveland, Ohio, 44112, U.S.A.)

The performance of a fluorescent lamp depends in part on the consistency of finished electrode coils. The room temperature resistance as well as the coil weight can be in error because of variability in tension during the coiling process. In addition, both can differ from desired values, even when processes are in control, if the equations used to predict and set process parameters do not adequately describe coiling. It will be shown in this paper that realistic values for these quantities can be calculated through use of equations based on existing knowledge of the deformation of metals.

While tension is a necessary aspect of the coiling process, high or variable tension usually is of paramount importance when weight control problems are encountered. In the past it was possible for weights of doubly wound coils to be as much as 15% lower than expected. Present day tensioning devices are still capable of excursions of several percent. Peculiarities in the manner of laydown of primary coiling on the secondary mandrel can result in less significant - but nevertheless measurable - coil weight differences. Combined effects of manner of laydown (1-2%) and full-to-empty bobbin differences (1%) can amount to a few percent variation in coil weight. The effect of laydown is difficult to include in coil calculations whereas the bobbin capacity effect can be readily included if desired.

Equations that describe coil dimensions in simple terms can be readily formulated. A departure from simple equations is required to describe the effect on weight, coil length, and cold resistance when the neutral axis, rather than the elastic axis, is used in the calculations. The use of the neutral axis results in predicted coil weights that are 2-3% lower and more realistic than when the elastic axis is used.

Low coil weights, short wire lengths, and low resistance can also be explained in part by the permanent indentation effects that occur as a result of high tension. Such effects can be produced at each of the coiling steps. Permanent effects can occur between side-by-side mandrels, on a mandrel resulting from an overwound wire, and during second and third coiling operations where both mandrel and overwind wires can be deformed. The net result of these indentation effects is an effective reduction in the mandrel diameter - leading, of course, to lower than expected coil weights.

The departure of coil dimensions from expected values can lead to variability in electrode performance in a number of ways. In lamp circuits where heating current is provided at constant voltage for the electrodes by means of taps on the secondary windings of the ballast, variation in the clamped wire length will result in a variation of the sustained average wire temperature. Early lamp end discoloration and reduced lamp life could result from a "properly" clamped coil that has a shorter than normal current carrying main wire.

The purpose of this work is to provide the means for better characterization and control of electrode coil behavior during manufacture as well as in normal operation. The study has resulted in more realistic computer programs for the initial design phase, better understanding and control of the decomposition of the carbonates during processing, and better final coil parameter control to give consistent electrode behavior during lamp starting and operation. An additional benefit of the more accurate calculation procedures is a more realistic appraisal of material utilization.

n° 57

MECHANISM OF GLASS DARKENING BY A LOW PRESSURE MERCURY DISCHARGE

B.J.Mulder and S. van Heusden

Philips' Research Laboratories

Eindhoven, The Netherlands

A study was made of the darkening of sodium and potassium silicate glasses by a low-pressure mercury discharge, using electron spectroscopy (ESCA), Rutherford Back Scattering (RBS) and Transmission Electron Microscopy as analytical tools. Initial changes seen with ESCA and RBS in sodium silicate glass exposed to a mercury discharge are twofold: The first observation is that mercury penetrates very rapidly from the discharge into the glass to depths of the order of 1 nm. It can be shown that excited mercury atoms enter the glass with an efficiency that is at least equal to the penetration efficiency of mercury ions when these strike the glass in the recombination process at the wall. A second observation is that sodium ions migrate away from the surface at a rate that is much lower than the penetration of mercury. The rate of the migration of sodium ions depends on the energy of the photons from the plasma. The migration is much enhanced when the intensity of strongly absorbed, energetic argon radiation is increased. Clearly, the sodium ions migrate in a photoelectric field. This field is caused mainly by trapping of photoelectrons at various electron traps in the glass, for instance, water molecules (OH-groups).

The simultaneous observation of the two processes: penetration of mercury into the glass and migration of sodium away from the surface might suggest that mercury ions simply replace alkali ions in the surface region of the glass. This is not, however, the case. Rather, the mercury penetrates into voids in the glass network that are sufficiently close to the surface. This follows from observations on potassium silicate glass exposed to a mercury discharge. In this case potassium ions do not disappear from the surface of the glass and nevertheless twice as much mercury is found at the surface (ESCA) than at the surface of the corresponding sodium glass. An explanation would be the larger molecular volume of potassium glass, implying more voids of sufficient size. From the "reservoir" of mercury in near-surface voids mercury migrates slowly deeper into the glass as mercury ions. This can explain that the migration of mercury is slower in mixed sodium/potassium glass than in single alkali glasses (mixed alkali effect on ionic mobility). The "mixed alkali" effect is especially large when the discharge current is low. This can be understood by assuming that the migration of mercury ions is not purely ionic, but is assisted, at higher discharge currents, by photoelectrons.

Ultimately the mercury ions are reduced by photoelectrons to atoms that agglomerate into droplets of metallic mercury responsible for glass darkening. These droplets were observed with Transmission Electron Microscopy. Once the nucleation of droplet's has started one observes a sudden increase in the rate of mercury take-up. The concentration of mercury in darkened glass peaks at depths around 30 nm.

The rate of glass darkening (involving penetration, migration, photoreduction, nucleation and growth of droplets) is vastly enhanced when the intensity of argon radiation from the discharge is increased by reducing the mercury pressure through a cold spot in the discharge tube. Argon radiation is more energetic than mercury radiation and much more strongly absorbed, producing a much stronger photoelectric field.

So far we have mentioned the role of excited mercury atoms and ions (penetration) and that of photons (photoelectric field, photon assisted migration, photoreduction of mercury ions). A fourth species supplied by the discharge and essential for the darkening process are low-energy electrons. These greatly enhance the strength of the photoelectric field in the glass, presumably by recombination with photoholes in the glass near the surface.

THE CHEMISTRY OF TIN HALIDE-SODIUM HALIDE SYSTEMS EMPLOYED IN HIGH INTENSITY DISCHARGE LAMPS

S.A. Mucklejohn and N.W. O'Brien
Research and Engineering Laboratories, THORN EMI Lighting Limited, Melton Road, Leicester, LE4 7PD, England.

The continued search for high intensity discharge lamps with high efficiency, long life and good colour rendition has given rise to an ever increasing array of metal halide lamps in recent years.

A mixture containing tin halides and sodium halides (Sn/Na/I/Cl/Hg) has been shown to be a particularly promising system for high intensity discharge lamps[1]. Previously, experimental work on silica bodied lamps containing the $SnCl_2$+NaCl system exhibited large enhancements in the sodium D line emission compared to lamps containing NaCl but no $SnCl_2$[2].

We have previously reported[3] the preliminary results of a Knudsen effusion/mass spectrometric study of the $SnCl_2$+NaCl, SnI_2+NaI, $SnCl_2$+NaI (all 1:1 molar ratio) systems. In this work we concentrated on measuring the contribution of sodium containing species to the total vapour pressure. All of these systems exhibit large enhancement in the sodium halide vapour pressure which is concordant with earlier lamp experiments. Although the total amount of sodium in the vapour above these melts is of great importance to lamp investigations, it is also relevant to consider the sodium containing species present at equilibrium and the vaporisation processes by which sodium is conveyed from the melt into the vapour.

We have subsequently measured the total vapour pressure above liquid melts of $SnCl_2$, SnI_2 together with the mixed metal halide systems: $SnCl_2$+NaCl, SnI_2+NaI, $SnCl_2$+NaI. The mixtures have been studied at various molar ratios up to temperatures found in the cool spot of lamps. All of these systems studied to date show large deviations from Raoult's

Law behaviour. This work has been complemented by further mass spectrometric studies of the $SnCl_2$+NaCl, SnI_2+NaI and $SnCl_2$+NaI systems.

Deviation from Raoult's Law behaviour in melts of metal halide systems suggests molecular interaction in the liquid. This is usually interpreted as the formation of 'complex' species; $NaSnCl_3$, $NaSnI_3$ for the SnX_2+NaX systems. Mass spectrometric studies of the SnI_2+NaI system[4,5] have recorded ions ($NaSnI^+$, $NaSnI_2^+$) derived from a complex molecule $NaSnI_3$. The ion-currents for species derived from the complex indicates a very low abundance for the complex in the vapour. Our previous work[3] showed no evidence for the presence of complex species above SnX_2+NaX melts, yet indicated large enhancements for the sodium halide vapour pressure. We interpreted this as due to rapid dissociation of the complex after evaporation.

The results of further studies on SnX_2+NaX systems will be described and the role of complex formation and the kinetic stability of these species will be discussed.

Mixtures of NaI and ScI_3 have been used in discharge lamps for 25 years, properties of the NaI+ScI_3 and SnX_2+NaX systems will be compared.

References

1. K.E. Brown, A.G. Chalmers and D.O. Wharmby, J. Illumin. Eng. Soc., 11, 114 (1982).
2. F.L. Whittaker, Lighting Res. and Tech., 13, 11 (1981).
3. S.A. Mucklejohn and N.W. O'Brien, Proc. 9th. Int. Mass Spec. Conf., Vienna (1982).
4. N.V. Karpenko and T.N. Sevast'yanova, Prob. Sovrem. Koordinat. Soed., 48 (1970).
5. K. Hilpert and K.A. Gingerich, Proc. 9th Int. Mass Spec. Conf., Vienna (1982).

n° 59

MASS SPECTROMETRIC STUDY ON THE THERMOCHEMISTRY OF THE VAPORIZATION AND OF THE GASEOUS SPECIES OF THE SYSTEMS $NaI-SnI_2$, $NaI-PbI_2$ AND $NaI-ScI_3$

K. Hilpert, H. Gerads, D. Kobertz, R. Viswanathan
Institute of Applied Physical Chemistry, Nuclear Research Centre (KFA)
P.O.B. 1913, D-5170 Juelich, Fed. Republic of Germany

Metal halide lamps contain NaI which vaporises at the wall of the hot silica tube ($T \sim 1000$ K) of such lamps. The molecular iodide vapours diffuse to the centre of the discharge in the arc where dissociation occurs on account of the high temperature ($T \sim 5500$ K). The released Na-atoms are excited and ionized resulting in an emission of light. The efficiency of the light generation can be enlarged if the density of the Na atoms in the discharge increases (ref. 1). This can be achieved by stable gaseous complexes which increase the amount of Na vaporised from the silica wall of the arc tube. In order to check the potential of SnI_2, PbI_2 and ScI_3 for the formation of such complex molecules and in order to obtain data for lamp modelling the vaporization of phases of the systems $NaI-SnI_2$, $NaI-PbI_2$ and $NaI-ScI_3$ was studied by high temperature mass spectrometry with a Knudsen cell.

The measurements were carried out with the Knudsen cell CH-5 mass spectrometer system supplied by Finnigan MAT, Bremen, W.-Germany. An ion counting system was mounted by us (ref. 2). In addition to our modification of the ion source power supply (ref. 2), the vacuum system and the mass selector were also altered (ref. 3). Knudsen cells machined from molybdenum (ref. 4) were used for the measurements. The knife-edged effusion orifices had a diameter of 0.3 or 1 mm. Temperatures were measured with a Chromel/Alumel-thermocouple (ref. 4).

NaI/SnI_2-mixtures of the molar compositions 2:1, 1:1 and 1:2 were investigated. The NaI/PbI_2- and NaI/ScI_3-mixtures investigated had the molar composition 1:1. In addition, the evaporation of the components NaI(3), SnI_2, PbI_2, and ScI_3 was studied seperately in order to facilitate the interpretation of the complex mass spectra obtained for the mixtures. These mass spectra contain the following ions: Na^+, I^+, Sn^+, NaI^+, Na_2I^+, SnI^+, SnI_2^+, $NaSnI_2^+$, and $NaSnI_3^+$ (system $NaI-SnI_2$)(5); Na^+, I^+, NaI^+, Na_2I^+, Pb^+, PbI^+, $NaPbI^+$, PbI_2^+, $NaPbI_2^+$, $NaPbI_3^+$, and $Pb_2I_3^+$ (system $NaI-PbI_2$); as well as Na^+, Sc^+, I^+, NaI^+, ScI^+, Na_2I^+, $NaScI^+$, I_2^+, ScI_2^+, $NaScI_2^+$, ScI_3^+, $NaScI_3^+$, $NaScI_4^+$, $Na_2ScI_4^+$, $Na_2ScI_5^+$,

$Na_3ScI_6^+$, and $Na_4ScI_7^+$ (sytem NaI-ScI_3). The dissociation enthalpies of the gaseous species $(NaI)_3$, $(SnI_2)_2$, $(PbI_2)_2$, $NaSnI_3$ and $NaPbI_3$ have been determined for the first time. Dissociation enthalpies of $(NaI)_2$, $(ScI_3)_2$ and $NaScI_4$ have been redetermined and better values were obtained. The value used so far for the dissociation enthalpy of $NaScI_4$ (6) was considerably higher than that expected by empirical rules. Vapour pressure enhancement factors for Na by complex formation are given.

Dissociation enthalpies were determined according to the second- and/or third-law method. The determination of enhancement factors was carried out by calibration with pure NaI.

References

1. D.C. Fromm, J. Seehawer, W.-J. Wagner: Lighting Res. and Technol. 8 (1976) 136

2. K. Hilpert, in: Advances in Mass Spectrometry, Vol. 7, (1977) 584, Proc. 7th Int. Mass Spectrometry Conf., Florence (Aug./Sept. 1976).

3. K. Hilpert: Ber. Bunsenges. Phys. Chem. 83 (1979) 161.

4. K. Hilpert: Habilitationsschrift Technische Hochschule Darmstadt, Darmstadt 1980; Jül-1744, Report from the Nuclear Research Centre (KFA) Jülich, Federal Republic of Germany, 1981.

5. K. Hilpert: Advances in Mass Spectrometry, Vol. 9, Proc. 9th Int. Mass Spectrometry Conf., Wien (Aug./Sept. 1982), in press.

6. D.E. Work: Lighting Res. and Technol. 13 (1981) 143.

n° 60

ELECTRODE TEMPERATURES IN HIGH PRESSURE GAS DISCHARGE LAMPS

F. Oostvogels and P. Tielemans

Nederlandse Philips Bedrijven B.V., Lighting Division

Eindhoven

The Netherlands

The temperature distribution of the electrodes in gas discharge lamps is important for chemical and physical transport processes, which affect lumen maintenance, lamp life and electrode performance. We measured temperature distributions under several conditions, i.e. work function, electrode dimensions and gas or vapour composition.

Work function

In lamps different systems are used with different work functions: bare tungsten, tungsten coated with thorium or barium oxide compounds. These systems have been studied in high pressure Ar and Hg discharges. We found with bare tungsten temperatures between 3000 and 3300K, corresponding to the work function of tungsten (4.5 eV). Thorium coated electrodes were found to operate at 2500-2800K which is in agreement with W-Th work function values of about 3.3 eV. Application of barium oxide compounds result in temperatures of 1800-2000K (work function 2.0 to 2.5 eV for BaO).

In metal halide lamps we found temperatures corresponding with the above values for bare tungsten. In lamps containing ThI_4 additive, we found 2700K. In a new type metal halide lamp with TmI_3, NaI and TlI additives we found with bare W electrodes lower temperatures than expected, namely 2800K. Probably the work function is reduced by thulium metal on the tungsten surface.

The above mentioned measurements for Ba-activated electrodes were carried out on mercury lamps. In high pressure sodium lamps, made of sapphire, the electrode temperatures are about 300K lower.

Geometry

In practice most electrodes are made from tungsten wire with a construction consisting of a rod and a coil. An important design parameter is the diameter of the wires. Because of the complex heat transport in coils, we choose rods as electrodes for model experiments, with the diameter as parameter. Temperature profiles of W and W-Th electrodes were measured as a function of the current (ac and dc). There exist two operating modes, small currents and thick rods show an unstable spot mode, high currents with relatively thin rods show a stable diffuse mode.

The temperature profile of the electrode can be discribed by a simple differential equation based on axial heat conduction and radial cooling by radiation. In the case of a diffuse mode, we found a good agreement between experiments and calculations based upon a simple model of electron emission. A simplified form of the heat transport equation shows simple relationships between the current (I) and rod diameter (d) for equivalent temperature distributions: $I/d^{3/2}$ =constant. In practical lamps with certain electrode configurations we found an agreement with this relationship.

Gas and vapour composition

In different rare gases we found relative small differences in electrode temperatures. This is attributed to variations of the ion current contribution and influence of the electric field on the electron emission, which depend on ion mass and ionisation potential. The lower temperatures in sodium vapour compared with mercury vapour is probably not due to this mechanism. Lowering of the work function is probably the main mechanism.

Gap in pagination due to unavailable paper.

Pages 156-157

The Scanning Electron Microscope as a Diagnostic Tool For High Pressure Sodium Lamps

Robert F. Craig
GTE Lighting Products
Sylvania Lighting Center
Danvers, MA 01923 USA

The scanning electron microscope (SEM) and energy dispersive x-ray spectrometer (EDS) lend themselves to the exploration and diagnosis of any number of problem areas in high intensity discharge devices. In this instance we will confine our discussion to the high pressure sodium (HPS) arc tube, and more particularly, to three areas of the lamp. These areas include the electrode emissive material, the polycrystalline alumina (PCA) arc tube and the "frit" material used to hermetically seal the lamp.

The application of the SEM-EDS technique to this type of lighting product allows an experienced investigator to explore micro features in the 1-2 um size range with little difficulty. Additionally, with the use of suitable standards, quantitative analysis of reasonable accuracy and precision can be obtained which can further enhance the understanding of the material dynamics in the aging process of HPS lamps. As will be shown, there is a wide range of dynamic chemistry taking place in the HPS device through the life of the lamp. Information relative to these changes is of primary importance to the materials scientist and lamp technologist for determining design parameters as a particular lamp type evolves toward maturity. Our principle concern in this work is to examine and evaluate the chemical and physical changes of the three mentioned component areas of the HPS lamp. These areas are the most vulnerable to change during life and are determining factors for lamp life.

The electron microscopic evaluation of electrode emissive material shows that routine processing as well as lamp aging can effect the chemistry of the barium calcium tungstate (BCT) emissive coating. These changes are manifest in a number of ways, from a simple change in the ratios of the elements involved to drastic alternation of the material. Such drastic change is seen as the formation of new compounds which include the addition of aluminum and loss of one or more starting elements. This type of reaction appears to involve the transport of either alumina or an aluminate from the arc tube wall or the sealing material. These reactions are particularly noticeable in lamps which have higher than normal sodium pressure.

Similarly the interior surface of the PCA arc tube is affected by the discharge plasma. The most common effect is the formation of beta alumina phases, thus altering both the chemistry and physical features of the arc tube surface. As with electrode reaction, the formation of beta alumina is dependant upon the sodium pressure as well as other considerations such as the initial purity of the starting materials used to form the PCA arc tube. Quite often the three major cations, Na, Ba and Ca are found in one microscopic feature. With the SEM-EDS the examination of an arc tube interior surface is simple and direct requiring minimal sampel preparation.

Our third and final area of concern is the ceramic "frit" used to seal the various parts of the lamp into a finish arc tube. Of particular interest in this area is the nature of crystallization of the material when the actual seal is made. In the system described here the "frit" is a calcium barium aluminate with minor additives. A major problem phase often forms, referred to as $C_{12}A7$, which appears to be detrimental to the seal's resistance to sodium attack. Examination of a seal with known starting composition and thermal history readily shows the physical features of phase assemblage, but also shows the chemical make up of each individual phase. This approach allows the lamp technologist to evaluate the material both for chemistry and the effects of sealing time and temperature. These same techniques are also applied to seal evaluation of operated lamps to determine the degree of sodium penetration and interaction in the lamp seal.

The application of the SEM-EDS technique to lighting devices offers a versatile and straightforward means of gaining information which by any other method would be difficult if in fact practical at all. As the methods and results are discussed, the appropriate sample preparation techniques will be presented.

n° 63

SECONDARY ION MASS SPECTROMETRY: A MATERIALS DIAGNOSTIC TECHNIQUE FOR THE LIGHTING INDUSTRY

D.E. Sykes, J.M. Walls and J.F. Raffle
Department of Physics, Loughborough University of Technology, Loughborough, Leics. LE11 3TU, U.K.

Secondary Ion Mass Spectrometry (SIMS) has developed into a powerful technique for surface analysis for all elements and isotopes. The importance of the technique lies in its remarkable sensitivity. It is capable of detecting most elements in the parts per million to parts per billion range. Recent advances in the instrumentation allow elemental (or isotopic) images to be obtained from an area 25μm to 400μm in diameter with sub-micron resolution. These combined with concentration-depth profiles obtained by simultaneous ion erosion and mass analysis enable high sensitivity, three-dimensional analyses to be carried out.

A thorough understanding of materials and processing can lead to significant improvements in production yields. In this paper the power of the SIMS technique is illustrated for materials characterisation and production problem-solving. Examples of its use include the identification of impurity elements in grain boundary segregation and void structures in tungsten and molybdenum; the compositional analysis of contaminating layers and the analysis of grain boundaries in alumina used for high pressure sodium tubes.

n° 64

THE PREPARATION OF VERY PURE LANTHANIDE IODIDES FOR METAL HALIDE LAMPS

W. Verbeek and J. Harmsen

Nederlandse Philips Bedrijven B.V. , Lighting Division

Eindhoven

The Netherlands

The quality of metal halide discharge lamps depends strongly on the purity of the metal iodides that are used. Contaminations such as hydrogen and water have a negative influence on physical and electrical lamp performance. Therefore the preparation of anhydrous metal iodides is of the utmost importance.

Metal iodides, especially the lanthanide iodides are very hygroscopic and hence difficult to synthesize with high purity. One of the known preparation methods starts from a hydrated salt, synthesized by dissolving the lanthanide oxide in freshly distilled hydrogen iodide. Dehydration is performed by heating in vacuum a mixture of the hydrated salt with an excess of ammonium iodide. After sublimation the salt should be pure enough to be used in metal halide lamps. The lighter lanthanide iodides, e.g. CeI_3, PrI_3, NdI_3 etc. have been prepared successfully in this way.

To prepare the heavier lanthanide iodids, e.g. ErI_3, TmI_3, LuJ_3 with a good yield and a high purity we had to modify the previous method. Dehydration is the most critical step in this preparation method. During this process the water molecules in the hydrated salt are replaced by ammonia molecules. It is important to remove the water molecules at low temperatures because at temperatures > 100^oC they can react with lanthanide iodide to form all sorts of unwanted products, e.g. lanthanide oxyiodide, lanthanide hydroxyiodide and iodine. We have succeeded in synthesizing most of the lanthanide iodides with a good yield and a high purity by employing a very slow heating process (6°C/hour, starting at 20°C) and a high vacuum (< 10^{-5} torr) during dehydration. Sumblimation is performed also under high vacuum conditions.

Lamps filled with these rather pure salts show excellent electrical behavior and good properties during life.

n° 65

DETERMINATION OF TRACE OXYGEN AND HYDROGEN IN HIGH-PURITY METAL HALIDES

Timothy R. Brumleve
Anderson Physics Laboratories, Inc.
406 N. Busey Ave.
Urbana, IL 61801 USA

This paper describes the application of coulometric Karl Fischer (KF) titration to the determination of trace quantities of O- and H-containing impurities in high-purity metal halide salts used as discharge lamp fill materials. In the KF titration trace H_2O is titrated with coulometrically generated I_2 in a solution of SO_2, pyridine, and methanol: $H_2O + I_2 + SO_2 \rightarrow 2HI + SO_3$. Metal hydroxides and some reactive metal oxides also undergo quantitative reaction in a similar fashion. The method is ideally suited to the analysis of metal iodides which are dissolved directly in the KF reagents. The sensitivity of the technique is excellent, giving useful results down to about 2 μg of H_2O (or down to 200 ppb for a 10 g sample.

The experimental setup is centered around a Mitsubishi CA-02 Moisture Meter and a special inert gas/vacuum line for sample introduction. The apparatus and principles of operation will be described in detail. Results of the analyses of some typical metal halide materials will be presented, and the effects of handling and storage on the measured H_2O/OH^- content will be discussed.

IMPROVED EFFICIENCIES FOR FLUORESCENT LAMPS BY ALTERING THE MERCURY ISOTOPIC DISTRIBUTION

S. G. Johnson, Dale E. Work*, Jakob Maya and John F. Waymouth
Sylvania Lighting Center
Danvers, Massachusetts

Experimental and theoretical studies of the imprisonment of resonance radiation within mercury vapour have been reported by a number of groups[1-3]. We will demonstrate, for the first time, that the efficiency of the plasma in a fluorescent lamp can be enhanced by altering the isotopic distribution of the mercury so as to decrease the imprisonment time of the resonance radiation at 254nm.

The imprisonment time of the resonance radiation is a consequence of the self absorption of the radiation by the ground state atoms within the plasma. The degree of absorption will be proportional to the number density of ground state atoms and the hyperfine structure of the atomic line.

The 254nm resonance line of mercury is composed of a series of hyperfine lines whose separations are an effect of the differences in the isotopic masses. Figure 1 shows the relative intensities and displacement of the hyperfine lines for the isotopic distribution occuring in natural mercury.

The line widths shown in Figure 1 represent the doppler width at half maximum for the isotopes at a temperature of 40°C. It is apparent from Figure 1 there is significant overlap of hyperfine lines for the odd isotopes with themselves and the even isotopes Hg^{198} and Hg^{204}. Consequently the hyperfine structure appears as a five line system. It is this structure and the relatively even distribution among the lines which permits the Hg plasma to be less self absorbing, and therefore more efficient, than a similar plasma emitting a resonant line of a less complex nature. It follows that by adding yet another component to the hyperfine structure the radiative lifetime would be further decreased and the plasma efficiency increased. This was accomplished by enriching the mercury with isotope Hg^{196} which, in natural mercury, is in low abundance, 0.146%, and which is well spaced from the other lines as shown in Figure 1.

Mercury, isotopically enriched in Hg^{196}, was obtained from a thermal reduction of mercuric oxide. The enriched mercury was combined with natural mercury to obtain a distribution of enriched mixtures which were dispensed from ampules into the lamps following processing. The initial test sampling contained three lamps with the Hg^{196} concentrations of 0.146% (natural mercury), 3.75%, and 15.3%. The lamps had the physical and operating characteristics of a 40 watt T-12 fluorescent lamp, except a central section of the lamp was made of quartz. Measurements of the relative intensity of the 254 nm radiation made through the quartz section showed the outputs of the 3.75% and 15.3% Hg^{196} lamps were 4.8% and 4.2% greater, respectively, that than of the control lamp when operated under constant current conditions. This enhancement of resonance radiation suggests an increase of at least 100 lumens in the radiant output of a T-12 four foot 40 watt cool white fluorescent lamp.

To verify this conclusion a series of 40 watt lamps were made on production equipment. The series consisted of nine test groups of between three and six test lamps, which represented a gradation of enriched mixtures from 0.146% Hg^{196} to 17.1% Hg^{196}. Photometric measurements showed the efficiency of the lamps increased with isotopic enrichment and that the greatest gains are achieved between one and six percent enrichment; the latter achieved a 3% increased in efficiency relative to the control. The results are in agreement with our theoretical predictions on lamp efficiency as a function of the resonance imprisonment time and isotopic distribution.

*present address: Westinghouse Electric Corp., Bloomfield, New Jersey

In conclusion, we have demonstrated an increase in the efficiency of the fluorescent lamp by modifying the hyperfine structure of 254 nm resonance radiation through an increase in the Hg196 concentration relative to the other naturally occurring isotopes. We believe that optimization of the mercury isotopic distribution will lead to a significant gain in the efficiency of all fluorescent lamps irrespective of geometry or phosphor.

Work continues in our laboratory to confirm and extend these results and to investigate the economic feasibility of isotopically-enriched mercury for fluorescent lamps. This continuary work is supported by the US DOE, Lawrence Berkeley Laboratory Subcontract No.451690.

1) T. Holstein, D. Alpert, and A. O. McCoubrey, Phys. Rev. 85, 985 (1952); T. Holstein, Phys. Rev. 83, 1159 (1951). T. Holstein, Phys. Rev. 72, 1212 (1947).

2) P. J. Walsh, Phys. Rev. 116, 511 (1959).

3) C. van Trigt, J. Phys. D 6, 1247 (1973); C. van Trigt, Phys. Rev. A 4, 1303 (1971); C. van Trigt, Phys. Rev. A 1, 1298 (1970); C. van Trigt, Phys. Rev. 181, 97 (1969).

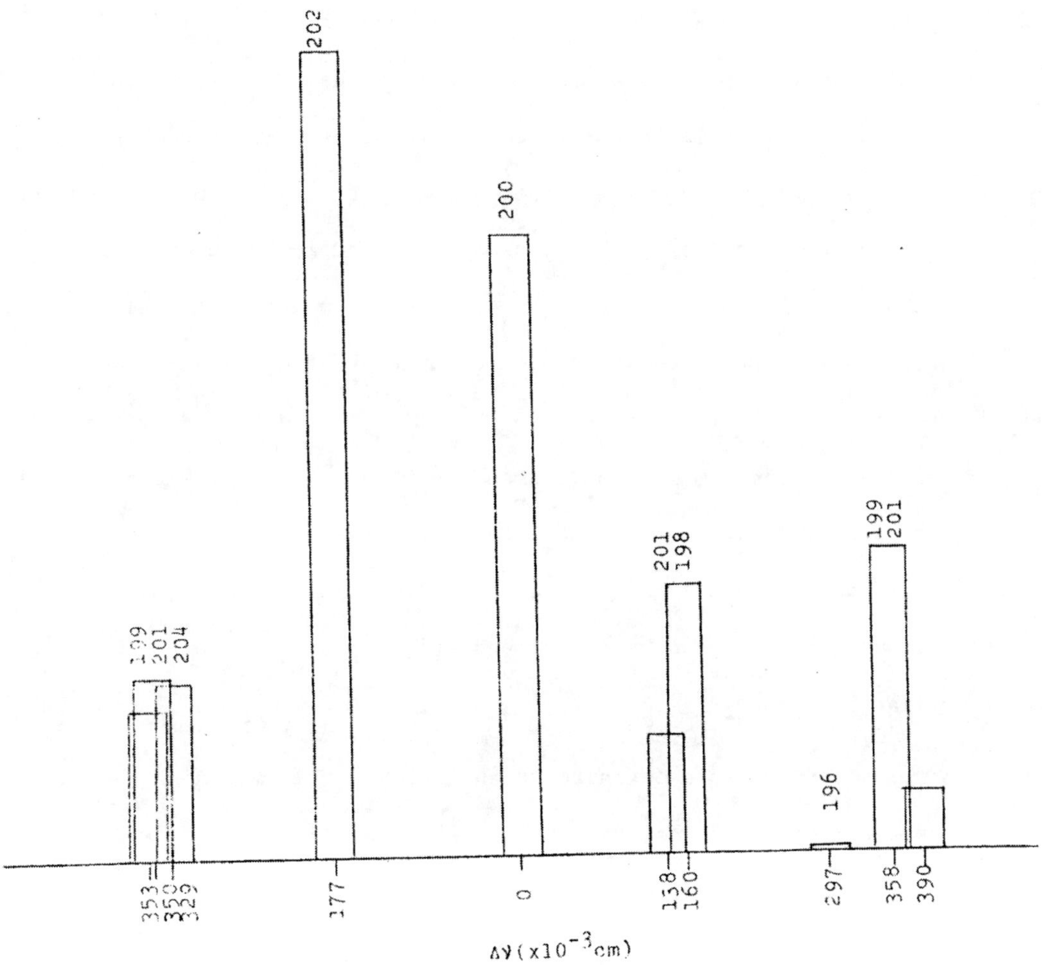

Hyperfine Components of the Mercury Resonance Line

Figure 1

n° 67

Theory of the Effects of Magnetic Fields and Isotope Enrichment on the Radiant Emittance of the Hg-Ar Discharge

by

R.W. Richardson*
and
S.M. Berman

Lawrence Berkeley Laboratory
University of California
Lighting Systems Research

Abstract

A variational principle for the radiant emittance of a Hg-Ar discharge is used to study the effects of magnetic fields and changes in the isotopic composition of the Hg. The principle is based upon time independent description of the discharge which uses the radiation transport equation to represent the photons and the diffusion equation for the Hg atoms in the presence of electrons which are characterized by a constant density and an energy distribution characterized by an electron temperature. We show that the emittance of such a plasma has a complex dependence on the strength of an external magnetic field. In particular, we show that for a single, even isotope of Hg, the emittance goes through a maximum when the Zeeman-effect splitting of the excited state is about 1.7 times the thermal line-width. The predicted value of the emittance at this maximum is 5.7 times its value at zero field and 1.9 times its value in the strong field limit. The dependence of the emittance on the isotopic composition of the Hg in the absence of a magnetic field is also discussed. Here, 3 of the 10 hyperfine lines behave independently while the other 7 are coupled together by line overlaps.

We determine that there is a definite enhancement in the emittance over that which occurs in natural mercury when the composition is modified to produce the optimum. The results and method of calculation are presented.

*Permanent address:
New York University
Physics Department
4 Washington Place
Newport, NY 10003

n° 68

ELECTRON MOBILITY CALCULATIONS FOR MERCURY-ARGON LOW PRESSURE GAS DISCHARGES

Z.F. Chen* and H.G. Jones

Department of Physics, Loughborough University of Technology, Loughborough, Leicestershire LE11 3TU, U.K.

*Institute of Electric Light Sources, Fudan University, Shanghai, China.

The electron mobility in Argon and Mercury-Argon mixture as used in fluorescent lamps are studied here. The effect of varying the Mercury concentration is expressed as a correction, where value is determined over a range of working temperatures.

For a particular gas, the electron mobility can be calculated using

$$\mu_e = \frac{e}{3kT_e p_o} \int \frac{vf}{Q_m} dv \qquad \ldots (1)$$

where f is the velocity distribution function, p_o the reduced pressure T_e the electron temperature, and Q_m the momentum transfer cross section usually obtained from published data.

In the case of Argon Q_m can be expressed as

$$Q_m = 400 \left(\frac{mv^2}{2e}\right)^{1\cdot05} \qquad \ldots (2)$$

giving an integration

$$\mu_A = 1.148 \times 10^9 / T_e^{1\cdot55}\, p_{Ao} \qquad \ldots (3)$$

This new expression for μ_A gives values that are in good agreement with Verweij (1961). For the mixture, Q_m is the sum of that for Argon and Mercury respectively. Hence the electron mobility is expressed as

$$\mu_e = \frac{1.148 \times 10^9}{p_{Ao}\, T_e^{1\cdot55}} (1-A) \qquad \ldots (4)$$

Where A is the correction factor, which is zero for zero Mercury concentration and determined numerically from

$$A = 1.7612 \times 10^9 \frac{p_{Mo}}{p_{Ao}} \int \frac{x^{0\cdot9} \exp(-x^2)}{x^{3\cdot5}\, T_e^{1\cdot75} + 9.0828 \times 10^8 \frac{p_{Mo}}{p_{Ao}}} dx \qquad \ldots (5)$$

where $x = \left(\frac{mv^2}{2kT_e}\right)^{\frac{1}{2}}$.

For Argon filling pressure of 3 torr and 400ma current the correction factors are as follows:

Argon Temperature	20°C	30°C	42°C	50°C	70°C	80°C
Correction Factor A	0.152	0.232	0.370	0.474	0.704	0.792

The data shows that electron mobility in the mixture is dominated by Mercury at high working temperature.

The axial electron density in a Mercury-Argon discharge has been calculated in a manner similar to Waymouth (1971) but using equation (4) given here. The excellent agreement between measured and the calculated values, emphasise the importance of considering the combined effect of the gaseous mixture on the electron motion especially at high working temperatures.

THE IONIZATION COEFFICIENT IN NOBLE GAS-MERCURY MIXTURES

A.L.J. Burgmans, A.H.M. Smeets and M. Merks
Philips Research Labs., Eindhoven, The Netherlands

For a better understanding of the ignition of fluorescent lamps knowledge is needed of the ionization coefficient of the gases used[1]. Although fluorescent lamps have been around for over 40 years the ionization coefficients have never been measured, most likely because of expected experimental difficulties. For instance formation of amalgams in the apparatus and possible influence of impurities due to outgassing of the walls during the experiments which have to be taken at temperatures between 20 and 100°C. In our work these problems have been solved by using the proper materials and high vacuum techniques. As a result the total fraction of impurities is less than 10^{-5}. The experiments are made in a conventional parallel-plate apparatus[2], the current between the plates is measured at both constant electric field E and number density n, as a function of the plate separation. The cathode is illuminated by UV radiation to create the primary electrons. From these data the ionization coefficient η is derived; η is the average number of ionizations resulting from one electron moving through a potential difference of one volt. As an example the results for argon-mercury are given in the figure as a function of E/n for different values of the mercury atom fraction. Small admixtures of mercury cause a large increase in η in particular in the low E/n region due to Penning ionization

$$Ar^m + Hg \longrightarrow Ar + Hg^+ + e$$

and associative ionization

$$Ar^m + Hg \longrightarrow ArHg^+ + e.$$

At high E/n values direct ionization of the noble gas becomes the dominant ionization mechanism. For a given E/n value η increases with increasing value of the mercury atom fraction up to $n_{Hg}/n \simeq 6\times10^{-3}$. For higher values of n_{Hg}/n the ionization coefficient starts to decrease as a result of inelastic collisions of the electrons with the mercury. For argon-mercury and krypton-mercury mixtures, which have been investigated, η is found to be a function of n_{Hg}/n and E/n only which means that de-excitation of the noble gas by non-ionizing processes is proportional to the density of the noble gas. Then in the case of small admixtures of mercury the ionization coefficient can be written as[3] $\eta = \eta_{n.g.} + \eta_{mix}$ with $\eta_{n.g.} = k_{i,n.g.}n_{n.g.}/v_{dr}E$ and $\eta_{mix} = k_{e,n.g.}n_{n.g.}/v_{dr}E\left[1+\frac{k_{d,n.g.}\;n_{n.g.}}{k_{i,mix}\;n_{Hg}}\right]$.

From the experiments in pure noble gases one therefore obtains values for the rate coefficient for ionization of the noble gas, $k_{i,n.g.}$, and from the additional contribution to η in the mixtures one obtains the rate coefficient for excitation of the noble gas as well as the ratio of the rate coefficients for de-excitation of the noble gas by non-ionizing collisions and ionizing collisions respectively, $k_{d,n.g.}/k_{i,mix}$. Values for the drift velocity, v_{dr}, have been taken from the literature.

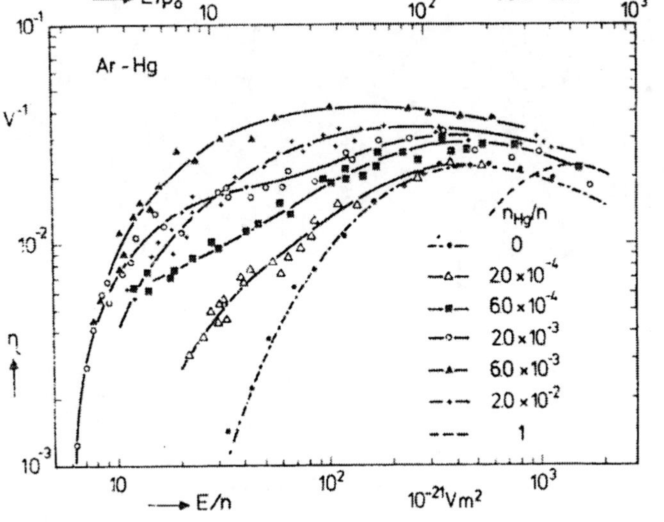

1. J.F. Waymouth, Electrical Discharge Lamps (M.I.T.) 1971.
2. J.M. Meeks and J.D. Craggs, Electrical Breakdown of Gases (John Wiley & Sons) 1978.
3. A.L.J. Burgmans and A.H.M. Smeets, to be published.

n° 70

ACOUSTIC RESONANCES IN HIGH FREQUENCY OPERATED LOW WATTAGE METAL HALIDE LAMPS

J.W. Denneman
Nederlandse Philips Bedrijven B.V., Lighting Division
Eindhoven
The Netherlands

The operation of high pressure gas discharge lamps at current frequencies higher than the mains frequency can be complicated by acoustic resonance phenomena.
These acoustic resonances can give rise to changes in the arc position or to an unstable arc at specific frequencies (ref. 1). In most metal halide lamps visual acoustic instabilities occur in broad frequency bands, indicating a large overlap of the resonance modes (ref. 2).

This paper describes how the resonance properties of small metal halide lamps were determined. This was achieved by operating the discharge lamps on a d.c. current with periodic polarity reversal to avoid cataphoretic effects. A high frequency sinusoïdal current is superimposed on the d.c. current. The frequency and amplitude of the superimposed current can be varied. By reducing the modulation depth of the high frequency power with respect to the total lamp power, the frequency ranges in which acoustic instabilities occur are reduced. In this way the resonance frequencies of the different modes can be determined as well as the minimum modulation depth needed to excite a visual instability.
The experiments have been carried out in spherically shaped lamps. The influences of the discharge tube radius, the mercury filling, the metal halide addition and the lamp power have been studied.

For a spherical lamp geometry, the acoustic resonance frequencies f_{ns} follow from the wave equation neglecting damping terms (ref. 3):

$$f_{ns} = \frac{\alpha_{ns} \cdot c}{2\pi R} \qquad (1),$$

with α_{ns} the sth zero of the first derivative of the spherical Bessel function j_n of the first kind and of order n. R is the sphere radius and c the effective velocity of sound. c is assumed to be constant across the volume of the lamp.

For the comparison with experiments, the value for c is obtained by inserting the experimental value for the lowest resonance frequency f_{11} into Eq(1). Using this effective c, the resonance frequencies for the higher order modes are calculated. The resulting values are in good agreement with the experimentally measured frequencies.

References:

1. H.L. Witting; J. Appl. Phys. 49 (1978), p2680.
2. J.M. Davenport; R.J. Petti; IES conference Atlanta (1982).
3. P.M. Morse, H. Feshbach; Methods of Theoretical Physics, Chapter 11, Mc Graw Hill, New York (1953).

n° 71

A novel approach to avoid acoustic instabilities in high frequency operated high pressure lamps

R. Schäfer
Philips GmbH Forschungslaboratorium Aachen, Germany

High frequency operation of discharge lamps has become increasingly interesting since the introduction of electronic power supplies. A serious problem associated here is the occurrence of arc instabilities in certain supply frequency regions, which greatly hinder proper lamp operation. This paper presents an approach which allows lamp operation at instability frequencies. As an introduction some considerations on arc instabilities associated with high frequency operation are given.

The common explanation for these arc instabilities is that the high frequency modulation of the electrical power input causes pressure fluctuations of the gas volume which can lead to standing pressure waves at particular power frequencies (acoustic resonances). The excitation of acoustic resonances in an adiabatic gas is described by the wave equation[1,2]

$$\frac{\partial^2 p}{\partial t^2} + \Gamma \frac{\partial p}{\partial t} - c^2 \Delta p = (\varkappa - 1) \frac{\partial N}{\partial t} \quad (1)$$

where p is the pressure fluctuation, N the power input per unit volume, Γ a damping constant, $\varkappa$ the ratio of specific heats at constant pressure and volume, respectively, and c the velocity of sound. For longitudinal resonances in a cylindrical arc vessel (length L, axial coordinate z) and harmonic power input the power can be expanded into a series

$$N(z,t) = \sum_{m=0}^{\infty} N_m \cos m\pi \frac{z}{L} \, e^{i\omega t} + \bar{N} \quad (2)$$

($\bar{N}$ being the average power). The solution for p is then given by

$$p(z,t) = \sum_{m=0}^{\infty} \frac{i\omega(\varkappa-1)}{\omega_m^2 - \omega^2 + i\omega\Gamma} N_m \cos m\pi \frac{z}{L} \, e^{i\omega t} \quad (3)$$

where $\omega_m^2 = (m\pi c/L)^2$ denotes the set of longitudinal resonance frequencies.

However, such standing pressure waves are not generally associated with arc instabilities. These acoustic resonances without visual instabilities are detected from the sound spectrum emitted by the lamp as well as from changes of the arc- and wall temperatures. The results show that a certain lower limit of the pressure amplitude is necessary to cause arc instabilities.

Therefore, stable lamp operation with pulsed supply voltages is possible for large duty cycles[3], since in this case the amplitudes N_m of the power input are small, thus causing only small pressure amplitudes according to eq. (3). A drawback is thatsimple power supplies for this type of operation are difficult to realize especiallywith respect to the necessity of lamp power stabilization. Another approach cited inthe patent literature[4] is sweeeping of the supply frequency, but in most cases, especially for low wattage lamps, arc instabilities cannot be avoided.

Here a new method to avoid acoustic instabilities is proposed, which can also be understood from eq. (3). For a resonance frequency $\omega = \omega_m$ the pressure is given by

$$p(z,t) = \frac{(\varkappa-1)}{\Gamma} N_m \cos m\pi \frac{z}{L} e^{i\omega_m t} \quad . \tag{4}$$

The latter equation shows that always a rigid phase relation exists between pressurefluctuation p and power input N. If this phase relation is disturbed during the build-up of the resonance strong resonant behaviour in the arc is hindered although it is operated at its resonance frequency. In our case we use a 180° phase jump of the power frequency generated by a 90° phase jump of the supply voltage and appearing every n-th period. With this supply mode e.g. a 300 W mercury lamp can be operated in its fundamental longitudinal resonance band (at about 6 KHz power frequency) without the occurence of arc instabilities. For this lamp the repetition of the phase jump could be altered between 0.5 and 10 periods of the supply voltage.The magnitude of the phase jump was varied between 50° and 130°. This mode of stablehigh frequency lamp operation can also be applied to other types of high pressure lamps as e.g. low wattage metal halide lamps which exhibit particularly broad instability regimes.

1 M. Schulz and U. Ingard, Phys. Fluids 10, 1031 (1967)

2 R. Schäfer and H.P. Stormberg, J. Appl. Phys. 53, 3476 (1982)

3 H.P. Stormberg and R. Schäfer, submitted to Light. Res. and Tech.

4 cf. US-PS 3890537 (1975)

n° 72

HOLOGRAPHIC INTERFEROMETRY APPLICABLE TO INVESTIGATION OF TEMPERATURE FIELD AND THERMAL LOSS IN DISCHARGE LAMPS.

Vdovin V.G., Pustoshkin A.A., Vdovina N.A., Pritkov A.A.

Thermal transfer processes play the important role in gas-discharge light sources with metal-halide additives (MHL), they determine their efficiency and energo-economy. Therefore the study of lamp thermal fields is one of the principal problems posed before elaborators.

In a series of works(1,2) applicable to isochoric (striking in closed region) high pressure arcs the authors developed the method of spacial distribution of determination the buffer gas temperature Ta (r, z) by holographic interferometry. This method consists in getting the interferogram for a lamp being in two states by means of double exposure holography and its decoding with allowance for radial (by r) and axial (by z) particle redistribution in the discharge. Finding Hg atom concentration distribution and knowing the temperature in any discharge point (for example on the inner wall or on the electrode) one can determine the atomic temperature in a discharge by means of Dalton relation. Holography application enables to get an interferogram of the discharge limited by the geterogenetic translucent wall without using special optical windows; the differentiality of this method allows to exclude the secondary process effect. Multicomponent composition of metal-halide plasma doesn't effect the interference kind of a picture; the evaluation shows that the contribution to Hg atom plasma polarizability is two orders higher the other component contributions. Since the interference bands while passing over the wall undergo the fracture, a zero band from which you can count a shift, in order to conserve a full number of particles is within the region(1). The value of the datum temperature is measured independently: by thermocouple- on the wall, by pyrometer- on the electrode. In determining Ta the error does not exceed 10% (2).

By means of suggested method there studied the thermal fields in radial and axial directions in discharges of two series lamp, discharge mercury 700 type with different thallium iodide additive (0,2, 4,6,7,9mg) and with the same additive of other components: NaI-20mg, InI-1,2mg, Hg-74mg; besides, the lamps of one series were filled by Xe with partial pressure of 300mm Hg, and also in mercury analog-metal-halide lamp, contains only Hg-80mg. All the lamps were studied in different loading conditions (300-800). The figure below showed typical distribution of atomic temperature in the middle discharge section for purely mercury lamp(1), for metal-halide lamp with 7mg TlI(2) and metal-halide lamp with 7mg TlI and Xe(3) at 700W. Based on obtained data there determined temperature gradients in different discharge areas and evaluated thermal loss.

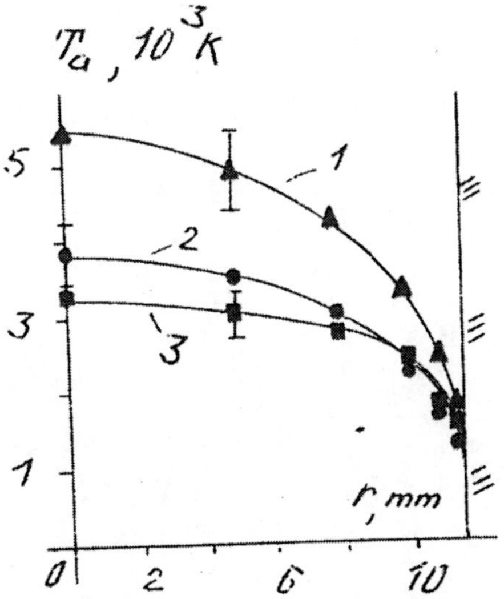

The study showed that in estimated limits of burning condition the temperature in the discharge centre remains constant and somewhat decreases with TlI dosage increased. The estimated temperature in the discharge centre is 1000-1500K lower the temperature determined by absolute brightness of Hg lines amounted 577/579nm which in a literature is assumed as a thermodynamic temperature. It confirms the finding that non-equilibrium takes place in metal-halide lamp discharge. The comparison of longitudinal and transverse gradients for purely mercury and metal-halide discharges showed that thermal loss in the last example is 30% lower.

Reference

1. Vdovin V.G., Pustoshkin A.A.-Theses of Report at All-Union Conference on Metrology of Transient Processes.-VNIIOFI,M., 1978, p.70.

2. Vdovin V.G., Pustoshkin A.A.-Theses of Reports at All-Union Conference - "Photometry and its Metrology Provision", VNIIOFI, M., 1979, p.235.

n° 73

USE OF THE 577 NM MERCURY LINE FOR THE DIAGNOSTIC OF HIGH PRESSURE LAMPS CONTAINING MERCURY

Jean-Jacques DAMELINCOURT, Michel AUBES, Adel ASSELMAN and Dominique GUILHEM
Centre de Physique Atomique, Laboratoire Associé au CNRS n° 277, Université Paul Sabatier, 118 route de Narbonne, 31062 TOULOUSE CEDEX, France
Godefroy LEMINTIER
Centre Technique de la Compagnie des Lampes, 11 Quai de Dion Bouton, 92818 PUTEAUX, France.

The 577 nm mercury line is often visible in lamp containing mercury as an active or buffer gas. It may be interesting to use the broadening of this line for the diagnostic of the discharge. This paper deals with the broadening of the 577 nm mercury line by mercury and electrons.

As an application abacuses are given for the determination of the mercury pressure in high pressure mercury lamps of 2 centimeters in diameter. If the mercury pressure is known one may estimate the electronic density. It is possible to extend this procedure to binary mixtures of mercury with other metallic vapors.

The 577 nm mercury line correspond upon a radiative transition between the 6^3D_2 and 6^1P_1 mercury levels.

The 6^1P_1 level is a resonant level and the main mechanism for neutral broadening is resonance broadening. Nevertheless, other broadening mechanisms have to be taken into account particularly electronic broadening. For resonance broadening we have followed the theory by H. R. Zaidi (1).

Zaidi considers resonance width for half intensity is proportional to $n_o(1 - n_o\beta)$ β is a parameter which does not depends upon neutral density or temperature.

For electronic broadening the work of M. A. Mazing has been used. The experimental determination of the broadening constants have been made by comparing experimental profiles with those obtained from the solution of the integral form of the radiation transfer equation.

The experiments and calculations have been performed for four discharges (one to five atmospheres). A.C. conditions permit the determination of the effect of atomic and electronic densities on the line formation.

The experimental values obtained for β and C_4 are :

$$\beta = 7\ 10^{-21}\ cm^3 \quad ; \quad C_4 = 1.4\ 10^{-12}\ rad\ cm^4\ s^{-1}$$

With the obtained values of β and C_4 we have calculated the line width for mercury discharges assuming local thermodynamic equilibrium and a temperature profile given by : $T(r) = T_c - (T_c - T_w)(z/R)^n$ with $n = 2 - 2.5 - 3$.

The results are given in figure 1. The possibility of extension to binary mixture is also presented.

REFERENCES

1. M. A. MAZING, Transaction of the PN Lebedev Physics Institute, volume XV, part. 2.
2. H. R. ZAIDI, Can. J. Phys. 26, (1977).

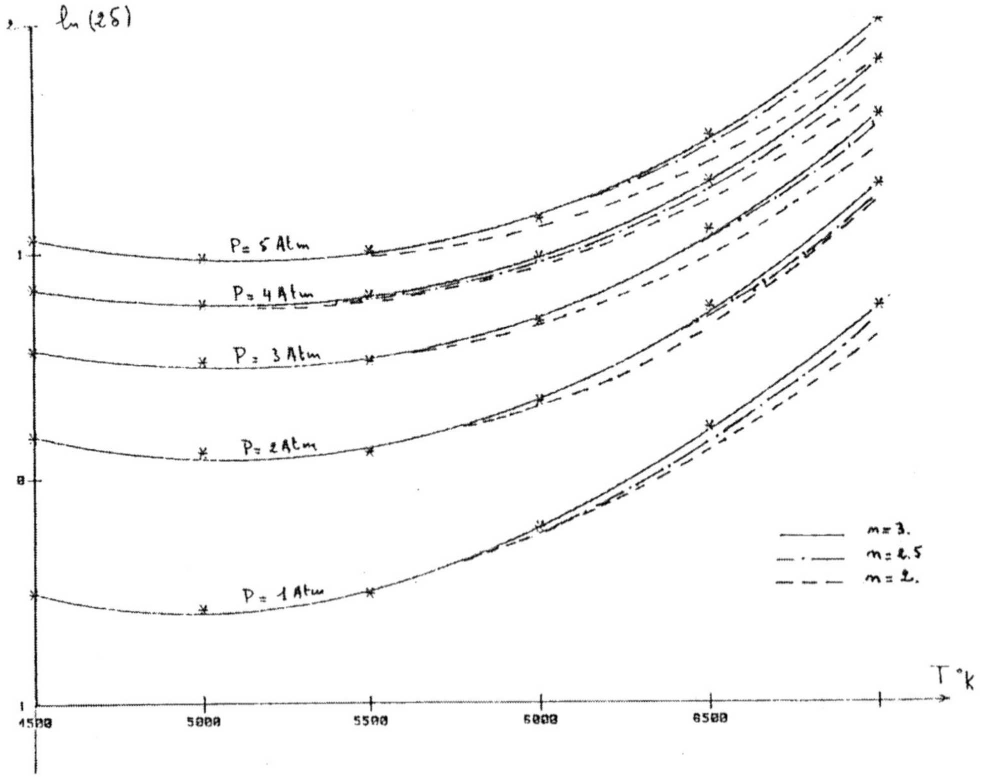

FIGURE 1 : Width of the 577 nm line for a diameter discharge of 2 cm.

n° 74

The radiation mechanism of the HMI-lamp

J.-Q.Lu* and H.-P.Popp
Lichttechnisches Inst.Univ.Karlsruhe,F.R.Germany
*Peking Institute of Light Sources,Hujialou,Peking

1.Introduction:

The HMI-lamp (OSRAM),which has high efficiency and very good color rendering, has found wide usage in the film and TV market. Because of its high wall-loading,the partial pressure of the rare-earth metal halides is high,compared with similar MHD-lamps. Early spectroscopic investigations showed a strong continuous radiation together with the common multi-line spectrum /1/.This paper presents results of quantitative measurements and evaluates the mechanism of this continuum.

2.The light source:

Instead of a commercial lamp, which contains different rare-earth metal halides,the investigated lamp is doped for simplicity with only Dy-metal halides. Dimensions and the other fillings are the same as with the 575 W HMI-lamp, namely: Hg,Cs,I,Br and Ar. Fig.1 shows some of the expected particle densities of 35 different species in this lamp, as a function of temperature at a total operating pressure of p=12 bar.

3.Experiment and results:

The lamp was operated vertically with constant input power of 550 W at 50 Hz. A 1m-monochromator with a resolution of 0,1Å was used for the time resolved measurements. The temperature profile was obtained by three different methods:

a) absolute intensity of the optical thin line Hg 577 nm.
b) peak intensity of the Hg 546 nm line (self reversal) applying BARTELs method.
c) reconstructing the self absorbed Hg 546 nm line profile according to BARTELs theory /2/.

The temperature of the arc core at current maximum reaches 6470 K.

The real Dy and I concentrations of the arc were determined from the optically thin lines Dy 562,7 nm and I 905,8 nm and are in good agreement with our calculations (Fig.1).

Side-on measurements of the continuous spectral emission and absorption were made at 7 line free "spectral windows" in the range 400 to 650 nm. By ABEL-inversion (including an absorption correction) the continuum emission coefficients are obtained, Fig.2.

In principle the continuum consists of two different parts.One is in the red range with a maximum at λ =610 nm, peaking at T=4300 K, while the other is in the blue range with a maximum at λ =420 nm, peaking at T=6000 K. We observed the same spectral emission components in a low pressure electrodeless microwave discharge containing Ar and DyI_3.

4.Evaluation and discussion:

With the particle densities (Fig.1) and the measured temperatures the KRAMERS-UNSÖLD-continuum (electron-ion recombination) and the free-free-minus continuum (Bremsstrahlung in the atomic field) were calculated.Their contribution is less than 6% of the measured continuum.

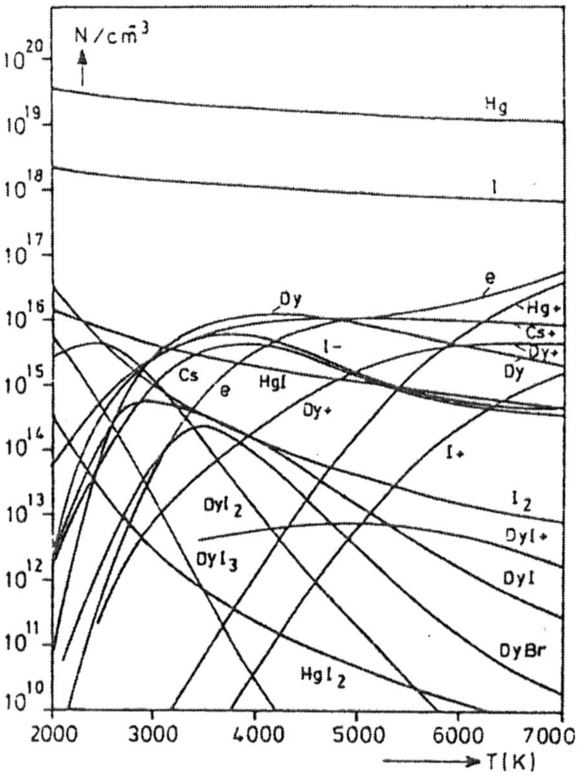

Fig.1:Calculated particle densities.Total pressure p=12 bar.

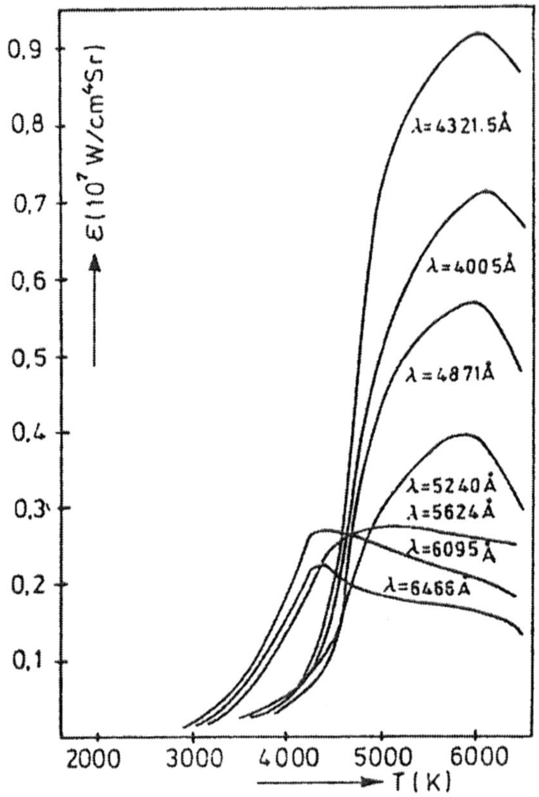

Fig.2:Measured emission coefficients of the Dy-HMI-lamp.

To identify the origin of the red and blue continuous radiation, we used the BOLTZMANN-plot technique together with the particle densities. Both continua can only result from merging band-spectra of the di-atomic DyI- / DyBr-molecules with excitation energies of 4.6 eV for the red and 7.8 eV for the blue part. This is in best agreement with our microwave measurements, where we observed a simple di-atomic band structure at very low densities. This is also supported by KRAMERs /3/ measurements on microwave discharges in Hg, LuI_3, CsI, Ar who identified the molecular emission as LuI-bands. Similar molecular spectra of LuI and DyI / DyBr were observed in a high pressure deuterium-discharge /4/.
The essential continuous part of the HMI-lamp radiation is therefore attributed to the di-atomic rare-earth halide molecules. The HMI-lamp is a molecular-lamp.

5.References:

/1/ H.Albrecht et al.:"Optische Strahlungsquellen",Lexika Verlag, Grafenau/Württ. 1977 S.194
/2/ D.C.Fromm et al.: Proc.Symp.High Temp.Chem.,Atlanta,1977 S.61
/3/ J.Kramer: J.Chem.Phys. **69**,2601 (1978)
/4/ W.Pilz,H.-P.Popp:11.Int.Conf.Phen.Ion.Gases, Prague 1973 S.409

* AUTHORS ALPHABETIC LIST *

* LISTE ALPHABETIQUE DES AUTEURS *